Lecture Notes in Computer Science   8457

Commenced Publication in 1973
Founding and Former Series Editors:
Gerhard Goos, Juris Hartmanis, and Jan van Leeuwen

Lecture Notes in Computer Science 8457

Maria J. Blesa   Christian Blum
Stefan Voß (Eds.)

Hybrid
Metaheuristics

9th International Workshop, HM 2014
Hamburg, Germany, June 11-13, 2014
Proceedings

 Springer

Volume Editors

Maria J. Blesa
Universitat Politècnica de Catalunya
Departament Llenguatges i Sistemes Informàtics
Jordi Girona 1-3, edif. Omega
08034 Barcelona, Spain
E-mail: mjblesa@lsi.upc.edu

Christian Blum
IKERBASQUE and University of the Basque Country
Department of Computer Science and Artificial Intelligence
Paseo Manuel Lardizabal 1
20018 Donostia, Spain
E-mail: christian.blum@ehu.es

Stefan Voß
University of Hamburg
Institute of Information Systems
Von-Melle-Park 5
20146 Hamburg, Germany
E-mail: stefan.voss@uni-hamburg.de

ISSN 0302-9743   e-ISSN 1611-3349
ISBN 978-3-319-07643-0   e-ISBN 978-3-319-07644-7
DOI 10.1007/978-3-319-07644-7
Springer Cham Heidelberg New York Dordrecht London

Library of Congress Control Number: 2014939565

LNCS Sublibrary: SL 1 – Theoretical Computer Science and General Issues

Typesetting: Camera-ready by author, data conversion by Scientific Publishing Services, Chennai, India

Printed on acid-free paper

Springer is part of Springer Science+Business Media (www.springer.com)

Preface

The International Workshop on Hybrid Metaheuristics was established with the aim of providing researchers and scholars with a forum for discussing new ideas and research on metaheuristics and their integration with techniques typical of other fields. The papers accepted for the ninth edition confirm that such a combination is indeed effective and that several research areas can be put together. Slowly but surely, this process has been promoting productive dialogue among researchers with different expertise and eroding barriers between research areas.

The papers in this volume are a representative sample of current research in hybrid metaheuristics. There are a large number of papers demonstrating how metaheuristics can be integrated with integer linear programming and other operations research techniques. Most of these papers are not only a proof of concept, which can be valuable by itself, but also show that the hybrid techniques presented tackle difficult and relevant problems.

In keeping with the tradition of this workshop, special care was exercised in the review process: out of 22 submissions received, 14 papers were selected on the basis of reviews by the Program Committee members and evaluations by the program chairs. The corresponding acceptance rate is higher than in previous editions. However, this was mainly due to the high average quality of the submitted works. Reviews were in great depth: reviewers sought to provide authors with constructive suggestions for improvement. Special thanks are extended to the Program Committee members who devoted their time and effort.

The present selection of papers can be of interest not only to researchers working on integrating metaheuristics with other areas for solving both optimization and constraint satisfaction problems. We hope that those who participated in HM 2014 succeeded in making connections between their own specific research areas and others.

June 2014

Maria J. Blesa
Christian Blum
Stefan Voß

Organization

Organizing Committee

General Chair
Stefan Voss   University of Hamburg, Germany

Program Chair
Christian Blum   Ikerbasque and University of the Basque Country, Spain

Publication Chair
Maria J. Blesa   Universitat Politècnica de Catalunya, Spain

HM Steering Committee

Maria Blesa	Universitat Politècnica de Catalunya, Spain
Christian Blum	Ikerbasque and University of the Basque Country, Spain
Luca di Gaspero	University of Udine, Italy
Paola Festa	Università degli Studi di Napoli FEDERICO II, Italy
Günther Raidl	Vienna University of Technology, Austria
Michael Sampels	Université Libre de Bruxelles, Belgium

Program Committee

Mauro Birattari	IRIDIA - Université Libre de Bruxelles, Belgium
Maria J. Blesa	Universitat Politècnica de Catalunya, Spain
Luca Di Gaspero	University of Udine, Italy
Karl Doerner	Johannes Kepler University Linz, Austria
Talbi El-Ghazali	University of Lille, France
Andreas Ernst	CSIRO, Australia
Paola Festa	University of Napoli FEDERICO II, Italy

Haroldo Gambini Santos	Universidade Federal de Ouro Preto, Brazil
Andrea Lodi	University of Bologna, Italy
Manuel López-Ibáñez	IRIDIA - Université Libre de Bruxelles, Belgium
Samir Loudni	Université de Caen Basse-Normandie, France
Vittorio Maniezzo	University of Bologna, Italy
Bernd Meyer	Monash University, Australia
Martin Middendorf	University of Leipzig, Germany
Panos Pardalos	University of Florida, USA
Günther Raidl	Vienna University of Technology, Austria
Andrea Schaerf	University of Udine, Italy
Patrick Siarry	Université de Paris 12, France
Thomas Stuetzle	IRIDIA - Université Libre de Bruxelles, Belgium
Stefan Voss	University of Hamburg, Germany

Additional Reviewers

Christian Kloimüllner	Sohel Rahman
Tommaso Urli	

Table of Contents

A CP/LNS Approach for Multi-day Homecare Scheduling Problems

Luca Di Gaspero and Tommaso Urli

DIEGM – University of Udine
Via Delle Scienze, 206 - 33100 Udine, Italy
{luca.digaspero,tommaso.urli}@uniud.it

Abstract. Homecare, i.e., supportive care provided at the patients' homes, is established as a prevalent alternative to unnecessary hospitalization or institutional care (e.g., in a rest home or a nursing home). These activities are provided either by healthcare professional or by non-medical caregivers, depending on the patient's needs (e.g., medical care or just instrumental activities of daily living).

In this paper, we consider the problem of scheduling Homecare Activities, that is, determining the caregivers' daily tours and the schedules of the homecare service to patients. We present a Constraint Programming (CP) formulation of the problem and we propose a Large Neighborhood Search method built upon the CP formulation.

1 Introduction

Supportive care at patients' homes is a prevalent alternative to the classical forms of institutional care (e.g., rest homes, nursing homes, hospitals) because it will increase the patient's quality of life while being more cost effective. *Homecare*[1] activities are performed by caregivers who visit the patient's home, carry out their tasks and then travel to the next patient. This specific feature makes the labor organization of homecare activities different from the one arising in institutional care. In particular, the patients' visits could have specific temporal and/or operator requirements which might impose the simultaneous presence of different caregivers at a given place, thus requiring a coordination of the tours.

In this paper we consider the problem of scheduling homecare activities for a time horizon H consisting of h consecutive *days* ($H = \{0, \ldots, h-1\}$). On each day $d \in H$ we have to schedule a set $\mathcal{A}_d = \{0, \ldots, n_d - 1\}$ of *activities* located in a given geographic location (x_a, y_a) (i.e., the patient's home), with a duration d_a and a number m_a of needed caregivers. In addition, specific requirements on the time window $[\sigma_a, \epsilon_a]$ in which the service must be provided can be imposed.

Activities are performed by a set $\mathcal{E} = \{0, \ldots, E-1\}$ of *caregivers*. Each caregiver $e \in \mathcal{E}$ starts and ends his/her tour from a geographic location (x_e, y_e) and can work on a (possibly different) specific time window $[\sigma_{e,d}, \epsilon_{e,d}]$ on each

---

[1] We do not distinguish here between activity of daily living and medical forms of homecare, however the latter is also known as *Home Healthcare*.

M.J. Blesa, C. Blum, and S. Voß (Eds.): HM 2014, LNCS 8457, pp. 1–15, 2014.

day $d \in H$. Moreover, he/she cannot perform all the possible activities in $a \in \bigcup_{d \in H} \mathcal{A}_d$ but only those which are *compatible* with his/her skills. This is stated by a binary relation $\rho_{e,a}$, whose value is 1 if caregiver e is compatible with activity a. Finally, because of labor regulations, each caregiver e should work at least $\underline{t}_{e,d}$ hours on a given day or he/she should have a day off. Moreover, a caregiver e cannot work for more than k consecutive days and for more than $\bar{t}_{e,d}$ hours on each day. Moderate violations to this latter constraint are usually allowed, but they will count as a overtime work, which has to be penalized.

The problem consists of determining the daily routes of caregivers and the schedules of activities so that the traveling costs and the use of overtime work is minimized. This problem is genuinely hard, being a combination of *employee scheduling* (more precisely *rostering*) and *vehicle routing with time windows* on a multi-period horizon.

It is worth noticing that the basic problem formulation relies on the hypothesis of existence of perfect solutions: i.e., those for which all activities are assigned to a caregiver. Unfortunately, in practice it is not always possible to serve all activities with the available caregivers, therefore in a more realistic formulation we will allow (but highly penalize) solutions in which some activities are left unassigned. The formulation of the problem in this form is the result of the real-world requirements collected and provided to us by EasyStaff, a company specialized in software solutions to timetabling and scheduling problems.

In this paper, we first introduce a Constraint Programming (CP) model of the problem that is based on a vehicle routing formulation. Then we show how we can generally build effective Large Neighborhood Search methods upon this model. Finally, we show with an experimental evaluation on the proposed solution methods on a set of random instances that simulate the structure of real-world Homecare assignment problems.

1.1 Related Work

To the best of our knowledge the first approaches to the problem are due to [1] and [2]. In the first paper, a simple scheduling heuristic is employed, whereas in the latter a MIP model has been formulated. Besides these early works, a few other modeling and solution approaches have been proposed for the Homecare problem. An established way, is to look at the problem from a set-partitioning perspective with side constraints [3,4] and solve it through variants of ILP methods (branch-and-price). In particular, Rasmussen et al. [3] focus on different temporal constraints among activities, which generalize the concept of *synchronization* constraints that will be described in Section 2.2.

Other works deal with a variant of the problem that considers multimodal transportation (i.e., car or public transportation), which is solved by hybrid approaches. Bertels and Fahle [5] use a combination of linear programming, constraint programming, and metaheuristics in a flexible tool that will handle multiple variants of constraints. Rendl et al. [6] solve the problem with a hybrid approach that employs CP for generating a valid initial solution to the problem and improves it through a number of different metaheuristics.

Other recent approaches to the problem are due to Trautsamwieser and Hirsch [7], who also employ a VNS-based metaheuristics, and Allaoua et al. [8], who devise a matheuristic approach based on the decomposition of the rostering and the routing parts.

Our work differs from the previous literature mainly because in all the cited paper the authors consider a daily time horizon. To the best of our knowledge [9,10] are the only papers on this topic that considers a periodic problem and they tackle it with an adaptive LNS approach. The main difference w.r.t. these works is that the authors of these two papers focus on stable solutions w.r.t caregivers, i.e., each patient should preferably always be visited by a single caregiver in the week. Moreover the CP model is based on a scheduling perspective and the LNS scheme they use is finer-grained that ours, since they relax a small set of activities either randomly or based on the penalty contribution to the cost function, whereas we consider a coarse-grained multi-day neighborhood.

Another difference of the present work with previous ones is that in some of the cited papers the authors explicitly model personnel skills, and allow an activity to be performed only by a sufficiently skilled caregiver. In our model, instead, skills are not explicitly considered but they are modeled through incompatibilities between activities and caregivers.

2   A Constraint Model for Homecare Activity Scheduling

The constraint model is based on the classical Vehicle Routing Problem (VRP) CP model [11] with some adaptations to the special structure of our problem.

We present our model in stages. Let first consider a single day, disregard the temporal components, assume that all homecare activities require exactly one caregiver, and that all activities have to be scheduled. In this setting, we build a special directed graph G, which consists of the following kinds of nodes:

- the departing node for each caregiver $S = \{0, \ldots, E\}$,
- the activities that should be executed by the caregivers in the tour $R = \{E + 1, E + 2, \ldots, E+,_1, \ldots, E + 1 + (n_d + r_d)\}$,
- the ending nodes for each caregiver $T = \{E + 1 + (n_d + r_d) + 1, \ldots, 2(E + 1) + (n_d + r_d)\}$.

An edge between two nodes (a_1, a_2) is weighted with the distance of traveling from a_1 to a_2.

The graph G contains $2E + n_d$ nodes, where $E = |\mathcal{E}|$ is the number of caregivers and n_d is the number of activities on day d. This graph structure allows to define *successor* and *predecessor* variables for each node of the graph, which will represent the caregivers routes. By imposing that the successor of the ending node for a given caregiver e is set to be the departing node of the following caregiver $e + 1 \mod E$, the problem consists in finding a Hamiltonian circuit in this graph. In this encoding, the sub-tour for the single caregiver e starts at node e (denoted by $\underline{e}$) and ends at node $E + n_d + e$ (denoted by $\overline{e}$).

We first remove the last assumption, i.e., we allow some activities to remain unscheduled. In order to do this, we use an additional *"dummy"* caregiver E, whose sub-path comprises all the activities that are left unserved. Consequently, the set of caregivers becomes $\mathcal{E} = \{0, \ldots, E\}$ and we will denote the set of "*regular*" caregivers by $\mathcal{E}^* = \{0, \ldots, E - 1\}$.

Secondly, we can easily remove the restriction on the number of caregivers required by an activity a by adding to the graph one replicate of the activity node for each caregiver required. As a consequence, the graph will have $n_d + r_d$ inner nodes, where r_d is the overall number of activity replicates on day d.

Thirdly, we introduce the time components by extending each node of the graph with a time window. As for the caregiver departing and ending node, the time window will coincide with the allowed worktime of the caregiver, whereas for the activity nodes it will consider the patient's temporal constraints.

Finally, the extension to multiple days is straightforward, as it will consider a set of distinct graphs, one for each day in the time horizon.

In the following, we give a detailed description of the variables, constraints, and cost function involved in our model. Moreover, we describe a custom branching strategy which exploits problem-specific knowledge to guide the search.

2.1 Variables

Similarly to the previous section, for the purpose of easing notation, the model variables are presented for a single day. The extension to multiple days is straightforward and will be denoted by using an additional superscript index $d \in H$ on all the following variables (which are summarized in Table 1).

As already mentioned, the routes in the graph G are represented by defining the successor of each node in V. Thus, we have $|V|$ successor variables succ that range over V, where succ_i represents the node following node i in the route. In addition, we define the redundant set of predecessor variables pred where pred_i denotes the node which comes just before node i in the route. Although these variables are redundant, according to [11], when channeled with the previous ones they achieve a more effective propagation.

Second, to each node i we associate the caregiver that will serve that activity[2] through the variable caregiver_i that ranges over $\mathcal{E}$.

The temporal components are captured by the variables start_i, duration_i, and slack_i, where the first two variables represent the start time and the duration of activity i, whereas the third is an optional waiting time of the caregiver after the activity has been served. In order to deal with manageable domain sizes for these variables, time values have been discretized by expressing them using timeslots of a given time granularity γ. For the purposes of this problem we choose $\gamma = 10$ minutes, which is adequate to express usual activity duration and practical traveling times, also allowing the possible compensation for small delays. The domain of the time variables for each day d, therefore, are proper subsets of the range $\{0, \ldots, 24\text{h}/\gamma\}$. Concerning the duration variables, in our problem they are statically determined in the problem instance we are considering. However, for

---

[2] We consider the starting and the ending nodes as placeholder activities that require exactly that specific caregiver.

Table 1. Variables in the CP Model; all variables are superscripted with the day d of the time horizon which they refer to

name[dimension]	domain	description
$\mathsf{succ}^d[V]$	V	successor of activity $i \in V$
$\mathsf{pred}^d[V]$	V	predecessor of activity $i \in V$
$\mathsf{caregiver}^d[V]$	$\mathcal{E}\nabla$	caregiver serving activity $i \in V$
$\mathsf{start}^d[V]$	$[0 \ldots 24\mathrm{h}/\gamma]$	the starting time of serving activity $i \in V$
$\mathsf{duration}^d[V]$	$\{0, d_i\}$	the duration of activity $i \in V$
$\mathsf{slack}^d[V]$	$[0 \ldots 24\mathrm{h}/\gamma]$	possible slack time after serving activity $i \in V$
$\mathsf{end}^d[V]$	$[0 \ldots 24\mathrm{h}/\gamma]$	the ending time of serving activity $i \in V$
$\mathsf{worktime}^d[\mathcal{E}]$	$[0 \ldots 24\mathrm{h}/\gamma]$	total worktime of caregiver $e \in \mathcal{E}$
$\mathsf{overtime}^d[\mathcal{E}]$	$[0 \ldots 24\mathrm{h}/\gamma]$	total overtime of caregiver $e \in \mathcal{E}$
$\mathsf{isWorking}^d[\mathcal{E}]$	$\{0, 1\}$	caregiver is working on this day
$\mathsf{unscheduled}^d$	$[0 \ldots n_d]$	number of unscheduled activities
$\mathsf{distance}^d$	$[0, ub]$	total traveled distance
cost^d	$[l, u]$	overall cost

modeling purposes we allow also the singular value 0 for duration that, along with the assignment to the dummy caregiver, will represent the duration of unscheduled activities. For this reason, they should be treated as variables. slack variables allow for flexible waiting times in case of spread activity time windows or activity synchronization. The end variables are just convenience variables used for representing the activity end time.

Finally, the variables worktime, overtime, and isWorking are defined on the set of regular caregivers $\mathcal{E}^*$. Variable $\mathsf{worktime}_e$ and $\mathsf{overtime}_e$ accounts for the daily worktime and overtime of caregiver $e \in \mathcal{E}^*$, whereas the boolean variables $\mathsf{isWorking}_e$ state whether the caregiver e is working or has a day off.

The components of the cost function we consider are the number of unscheduled activities, the total overtime (i.e., $\sum_{e \in \mathcal{E}} \mathsf{overtime}_e$) and the total traveled distance. These are aggregated in a linear combination and expressed in monetary units.

2.2   Constraints

The presentation of the constraints in our model are divided in two parts. First we introduce the essential constraints that are needed to thoroughly model the problem and then we discuss some redundant constraints that will make the tree-search process more efficient by allowing an early detection of infeasible solutions. As in the previous section we temporarily omit the day superscript d on the variables. The global constraints are presented using the GECODE syntax.

Essential Constraints. We start our description with the routing part of the model. The first set of constraints states the relationship between the successor and predecessor variables for all the nodes. This is achieved by channeling them through the `element` global constraint.

$$\texttt{element}(\texttt{pred}, \texttt{succ}_i) = i \quad \forall i \in V \tag{1}$$

$$\texttt{element}(\texttt{succ}, \texttt{pred}_i) = i \quad \forall i \in V \tag{2}$$

In order to build a circuit the connections between the ending node of one caregiver and the starting node of the following one are explicitly set.

$$\texttt{succ}_{E+(n_d+r_d)+e} = (e+1) \mod (E+1) \quad \forall e \in \mathcal{E} \tag{3}$$

$$\texttt{pred}_e = E + 1 + (n_d + r_d) + (e - 1 \mod (E+1)) \quad \forall e \in \mathcal{E} \tag{4}$$

Finally, the paths described by the `succ` and `pred` variables must be Hamiltonian circuits, and this is stated by means of the `circuit` global constraint. The variant of the constraint we use, also binds an array of variables with the distance of the edges selected by the path variables.

$$\texttt{circuit}(\texttt{succ}, \mathcal{TD}, forwardDistance) \tag{5}$$

$$\texttt{circuit}(\texttt{pred}, \mathcal{TD}, backwardDistance) \tag{6}$$

In the previous constraints, $\mathcal{TD}$ denotes the matrix with the traveling distances between any pair of nodes in the graph. This is a static data of the problem instance that can be either computed exactly (e.g., through GIS APIs) or just approximated (e.g., considering the haversine distance). Moreover, $forwardDistance$ and $backwardDistance$ are the arrays with the distances of the selected edges. These arrays cannot be simply summed up in order to compute the total traveled distance because also the path of the dummy caregiver is included in the circuit. We show in a while a workaround to this situation.

We proceed with constraints on the caregivers' variables. First, we initialize the correct caregiver $e \in \mathcal{E}$ for each of the starting and the ending nodes:

$$\texttt{caregiver}_{\underline{e}} = e \quad \forall \underline{e} \in \mathcal{E} \tag{7}$$

$$\texttt{caregiver}_{\overline{e}} = e \quad \forall \overline{e} \in \mathcal{E} \tag{8}$$

second, we push the caregiver-chain over the path variables on the regular activity nodes so that every node in a path must be served by the same caregiver:

$$\texttt{element}(\texttt{caregiver}, \texttt{succ}_i) = \texttt{caregiver}_i \quad \forall i \in R \tag{9}$$

$$\texttt{element}(\texttt{caregiver}, \texttt{pred}_i) = \texttt{caregiver}_i \quad \forall i \in R \tag{10}$$

third, we remove a given caregiver from the domain of his/her incompatible activities.

$$\rho_{i,e} = 0 \Rightarrow \texttt{caregiver}_i \neq e \quad \forall i \in R \tag{11}$$

As a result of these constraints, every activity in a path starting from node e and ending on node $E + 1 + (n_d + r_d) + e$ is served by the same caregiver e. Consequently, we can filter out the edges traveled by the dummy caregiver by means of a linear combination of the *forwardDistance* and *backwardDistance* variables whose multipliers are the boolean expressions stating whether the outgoing edge from node i is traveled by the dummy caregiver, that is:

$$\text{distance} = \sum_{i \in V} (\text{caregiver}_i = E) \cdot forwardDistance_i \tag{12}$$

Analogously, we can bind the number of activities that are left unscheduled with the following constraint:

$$\text{count}(\text{caregiver}, \{E\}, \text{unscheduled}) \tag{13}$$

Next, we present the temporal and scheduling constraints. However, before doing that let us describe some modeling assumption we made in order to have a tractable and uniform expression of these constraints. Since the number of activities that will be assigned to the dummy caregiver E cannot be predicted in advance, we decided to model the time variables of the special path starting from the depot E with "zero" times. That is, we write the constraints that impose all the start, the duration, and the slack to be 0 on that path. Conversely, on regular caregivers these variables are constrained according to the temporal constraints that are statically expressed in the problem instance.

Therefore, the first set of temporal constraints for regular activity nodes deals with the special case of the dummy caregiver:

$$\text{caregiver}_i = E \Rightarrow \text{start}_i = 0 \wedge \text{slack}_i = 0 \quad \forall i \in R \tag{14}$$

$$\text{caregiver}_i = E \iff \text{duration}_i = 0 \quad \forall i \in R \tag{15}$$

Conversely, for a regular caregiver the following constraints hold:

$$\text{caregiver}_i \neq E \Rightarrow \text{start}_i \geq \sigma_i \wedge \text{end}_i \leq \epsilon_i \quad \forall i \in V \tag{16}$$

These constraints are also extended on starting and ending nodes of the regular caregivers, where the time window $[\sigma_e, \epsilon_e]$ for those nodes is the working time window for the caregiver $e \neq E$. The variables end are used here to represent the ending time of the activities and they are constrained as follows:

$$\text{end}_i = \text{start}_i + \text{duration}_i + \text{slack}_i \quad \forall i \in V \tag{17}$$

For starting and ending nodes (S and T), the variables duration and slack are set to zero:

$$\text{duration}_i = 0 \wedge \text{slack}_i = 0 \quad \forall i \in S \cup T \tag{18}$$

The time-chain on the regular nodes is pushed through the following constraints:

$$\text{element}(\text{start}, \text{succ}_i) = \text{start}_i + \tag{19}$$
$$(\text{caregiver}_i \neq E) \cdot (\text{duration}_i + \text{slack}_i +$$
$$\text{element}(\mathcal{TT}, i, \text{succ}_i)) \quad \forall i \in S \cup R$$

where $\mathcal{TT}$ is the matrix of the traveling time between two nodes and the expression (caregiver$_i \neq E$) is a boolean multiplier that propagates the zero start in the case of the dummy caregiver.

Because of the modeling choice on activities requiring multiple caregivers, which are exploded in different replicates of the same activity node, we must ensure that those replicas are synchronized. Moreover, we require that either all the replicates are assigned to a suitable regular caregiver or all replicates remains unscheduled. These conditions are stated by the following constraints:

$$\texttt{count}(\texttt{caregiver}_{i,\dots,i+m_i-1}, \mathcal{E}^*, regularCaregivers) \quad \forall i \in R^* \tag{20}$$

$$\texttt{nvalues}(\texttt{start}_{i,\dots,i+m_i-1}, 1) \quad \forall i \in R^* \tag{21}$$

$$\texttt{nvalues}(\texttt{duration}i, \dots, i+m_i-1, 1) \quad \forall i \in R^* \tag{22}$$

The set R^* denotes the set of nodes that are the first replicate of an original activity and m_i is the number of caregivers needed. Variable $regularCaregivers$ is bound with the number of regular caregivers assigned to the activity replicates and its domain is $\{0, m_i\}$ so that is either zero or the correct number of replicates. The following two set of constraints impose synchronization of activities start times and duration.

Next, the schedule of the activities assigned to the same regular caregiver must not overlap in time. To this aim we consider each caregiver as a unary resource and the activities as tasks and impose the following constraints:

$$\texttt{unary}(\texttt{start}_{i\in R}, \texttt{end}_{i\in R}, (\texttt{caregiver}_{i\in R} = e)) \quad \forall e \in \mathcal{E}^* \tag{23}$$

The expression (caregiver$_{i\in R} = e$) is expanded in an array of $|R|$ boolean variables that activate the no-overlap scheduling constraint only on the indexes of the temporal variables for which the expression is true. These constraints can be efficiently propagated using a sophisticated sequence of methods (see [12]).

Finally, for this family of constraints, the computation of worktime and overtime is given by the following constraints:

$$\texttt{worktime}_e = \texttt{start}_{E+(n_d+r_d)+e} - \texttt{start}_e \quad \forall e \in \mathcal{E}^* \tag{24}$$

$$\texttt{overtime}_e = \max\{0, \texttt{worktime}_e - \bar{t}_e\} \quad \forall e \in \mathcal{E}^* \tag{25}$$

Because of the labor regulations concerning the daily worktime, we also impose a minimum amount of worktime $\underline{t}_e$ for caregiver e or we give him/her a day off:

$$\texttt{worktime}_e \geq \underline{t}_e \vee \texttt{worktime}_e = 0 \quad \forall e \in \mathcal{E}^* \tag{26}$$

The last family of constraints deals with the multi-day horizon of our problem. To this aim, we reintroduce from now on the superscript d on variables belonging to different days. The first constraint binds the isWorking variables:

$$\texttt{isWorking}_e^d \iff \texttt{worktime}_e^d > 0 \quad \forall e \in \mathcal{E}^* \tag{27}$$

According to these variables, the number of consecutive days a caregiver could work can be limited to k by the following sequence constraints:

$$\texttt{sequence}(\texttt{isWorking}_e^{d\in H}, 1, h, wd) \quad \forall e \in \mathcal{E}^* \tag{28}$$

where wd is a variable whose domain is $\{0, 1, \ldots, k\}$ thus allowing consecutive sequences of at most k ones in the isWorking variables in the time horizon H.

This concludes the description of the essential of our CP model for the home-care problem. The model can be enhanced by some redundant constraints, that will take care of some particular substructure of the problem and will be presented in the next section together with a custom propagator that will perform a look-ahead so to increase the pruning capabilities for the routing part.

Redundant Constraints. The first family of redundant constraints avoid sub-tours from a starting node to another starting node or from an ending node to another ending node.

$$\text{succ}_i \geq |S| \quad \forall i \in S \cup R \tag{29}$$

$$\text{pred}_i < |S \cup R| \quad \forall i \in S \cup R \tag{30}$$

Moreover, from starting or ending nodes we allow either a path through regular nodes or a short-circuit from a caregiver's starting node to his/her ending node:

$$\text{succ}_e < |S \cup R| \vee \text{succ}_e = E + 1 + (n_d + r_d) + e \quad \forall e \in \mathcal{E}^* \tag{31}$$

$$\text{pred}_{E+1+(n_d+r_d)+e} < |S \cup R| \vee \text{pred}_{E+1+(n_d+r_d)+e} = e \quad \forall e \in \mathcal{E}^* \tag{32}$$

Similarly to constraint (12), we impose that the computed traveled distance in should be also equal to the backward (i.e., pred) Hamiltonian circuit:

$$\text{distance} = \sum_{i \in V} (\text{caregiver}_i = E) \cdot backwardDistance_i \tag{33}$$

It is possible to filter out edges that cannot belong to any route either by looking at some static problem instance information or at dynamic assignments. As for the static filtering we look at the allowed time windows of pair of activities:

$$\text{caregiver}_i \neq E \wedge [\sigma_i, \epsilon_i] \succ [\sigma_j, \epsilon_j] \Rightarrow \text{succ}_i \neq j \quad \forall i, j \in V \tag{34}$$

The $\succ$ relation between time intervals is true when the left-hand interval do not overlap with the right-hand one and $\sigma_i > \epsilon_j$. The dynamic filtering also looks at pair of activities and their temporal assignments:

$$\text{caregiver}_i \neq E \wedge \text{start}_i > \text{end}_j \Rightarrow \text{succ}_i \neq j \quad \forall i, j \in V \tag{35}$$

We state the following synchronization constraint among replicates:

$$regularCaregivers = 0 \iff \text{caregiver}_i = E \wedge \text{caregiver}_{i+1} = E \wedge \ldots$$
$$\ldots \wedge \text{caregiver}_{i+m_i} = E \quad \forall i \in R^* \tag{36}$$

where $regularCaregivers$ is the one that corresponds to constraint (20).

Thanks to the isWorking variables, we can explicit pruning caregiver's values from the domain of regular activities, i.e.:

$$\neg \text{isWorking}_e \Rightarrow \text{caregiver}_i \neq e \quad \forall e \in \mathcal{E}^* \quad \forall i \in R \tag{37}$$

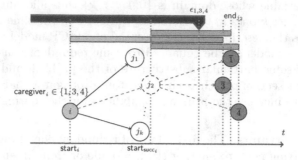

Fig. 1. Value j_2 can be removed from the domain of succ_i if all two steps paths from i to an ending node $\bar{e}$ passing through j_2 exceed the specific time window

In addition, we can immediately force the short-circuiting of the paths of non working caregivers:

$$\neg\mathsf{isWorking}_e \Rightarrow \mathsf{succ}_e = E + (n_d + r_d) + e \quad \forall e \in \mathcal{E}^* \tag{38}$$

$$\neg\mathsf{isWorking}_e \Rightarrow \mathsf{pred}_{E+(n_d+r_d)+e} = e \quad \forall e \in \mathcal{E}^* \tag{39}$$

and impose a conventional starting time to the departing and ending nodes so to break symmetries:

$$\neg\mathsf{isWorking}_e \iff \mathsf{start}_e = \sigma_e \wedge \mathsf{start}_{E+1+(n_d+r_d)+e} = \sigma_e \quad \forall e \in \mathcal{E}^* \tag{40}$$

Look-ahead propagator. The final redundant constraint we impose is a look-ahead constraint. This constraint has been implemented by means of a custom propagator that performs a one-step look-ahead of the temporal variables of the successors of a given node. The idea of this propagator (see Figure 1) is to prune a value j from the succ_i variable if all two-step-paths from i to the ending depot d_e passing through j will violate the temporal constraints (16). The addition of this constraint was motivated by preliminary experiments on the model.

Cost Function. The cost function of the problem is a hierarchical one, and comprises the three components measured by the variables unscheduled, distance, and overtime. Moreover, since we would like to maximize the efficient use of caregivers' worktime we also penalize the use slack in solutions.

We consider a weighted aggregation of these components to obtain a measure in a common unit of measurement, i.e.:

$$\mathsf{cost} = w_1 \sum_{d \in H} \mathsf{unscheduled}^d + w_2 \sum_{d \in H} \sum_{e \in \mathcal{E}^*} \mathsf{overtime}_e^d + \tag{41}$$

$$w_3 \sum_{d \in H} \sum_{i \in V^d} \mathsf{slack}_i^d + w_4 \sum_{d \in H} \mathsf{distance}^d$$

where $w_1 = 100$€, $w_2 = w_3 = 25$€/h, and $w_4 = 0.30$€/Km. These weight values have been set up according to the current work and traveling costs.

2.3 Branching Strategy

The branching strategy we employ is a bi-level one. First we determine the number of caregivers that will work on a given day of the timeslot. This is done through a surrogate counting variable that sums up the values of the isWorking boolean variables. The heuristic for value selection for this brancher is to take the median value. The motivation is that we have also to assign days off to caregivers, therefore it would be effective to spread these days off on the whole time horizon. Once the number of available caregivers have been set, a second brancher will randomly select the value of the isWorking variables accordingly.

In the second level of the branching strategy we aim at constructing the caregivers' routes and determine the activity schedule. To do so, in order to exploit the problem structure, we implemented a custom branching strategy, whose behavior is illustrated in Figure 2. The procedure tries to extend the shortest caregivers' route (Fig. 2(a)) by setting the succ variable of its current last step. The selection of the next activity is driven by the following heuristics:

1. prefer a regular node over an ending node;
2. if replicates of multiple activities have already been scheduled, prefer an unscheduled replicate of those activities;
3. prefer an activity that has to end earlier;
4. prefer shorter activities;
5. prefer an activity that has to start earlier;
6. prefer the activity with most replicates.

Once the succ variable has been selected, the brancher will set the start variable of the just added node to its earliest value (Fig. 2(b)). All the other temporal variables will be fixed thanks to constraint propagation. Once the routes of regular caregivers are closed (i.e., they reach the ending node), all the remaining activities are assigned to the dummy caregiver (Fig. 2(c)) and their temporal variables are assigned to the conventional value 0 (Fig. 2(d)).

3 Hybrid Approach

Large Neighborhood Search (LNS) [13,14] is a neighborhood search meta-heuristic based on the observation that exploring a *large neighborhood*, i.e., perturbing a significant portion of a solution, generally leads to optima of much *higher quality*. While this is an undoubted advantage in terms of search, exploring a large neighborhood structure can be *computationally impractical*, and requires a higher effort than exploring of a regular neighborhood. For this reason, LNS has been often coupled with filtering techniques, with the aim of reducing the size of the neighborhood by neglecting those choices that would lead to unfeasible solutions. In particular, LNS has been often associated with constraint models in order to tackle complex routing problems [15,16].

The LNS procedure is initialized by generating the first feasible solution using the branching strategy described above. This step is essentially equivalent

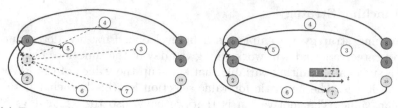

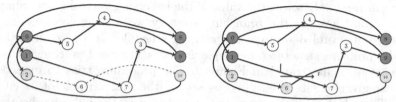

(a) Extend the shortest route by set- (b) Set the start variable of the activ-
ting the succ variable of its last node. ity just added to its earliest value.

(c) Assign remaining activities to the (d) Set their temporal variables to
dummy caregiver.   the conventional value 0.

Fig. 2. Custom tree-search branching strategy for the Homecare problem

to finding the first feasible solution with our CP model. Once the initial solution has been generated, the algorithm enters a refinement loop, which consists
of two alternate steps. First, the *destroy* step, relaxes (unassigns) a subset of
the decision variables, yielding a smaller constrained optimization problem (less
variables, filtered domains). Second, the *repair* step, re-optimizes the relaxed
variables by means of a tree-search.

In our destroy step, a fraction of the decision variables is relaxed uniformly
at random based on a parameter $\delta \in \{1, \ldots, h\}$, which represents the number
of full days that are relaxed by the destroy step, i.e., if $\delta = 1$ then all the
variables related to a single day are relaxed, if $\delta = h$ the solution is completely
relaxed. Since the pure CP tree-search assigns the variables day after day, in
this understanding this relaxation scheme allows to undo possibly bad decision
made by CP towards the root of the search tree.

Once a subset of the decision variables have been relaxed, a new solution is
produced through tree-search. Ideally, the repair step should return the best
solution in the neighborhood. However, depending on the number of relaxed
variables, finding the optimum can be non-viable. Our re-optimization strategy
consists in giving the repair step a time budget, which is proportional to the number of relaxed variables, i.e., a fixed number of milliseconds t_{var}. Moreover, this
tree-search chooses the number and identity of the working caregivers uniformly
at random. Our choice of relaxation and re-optimization steps is completely
neutral to the model constraints, including the sequence constraints (28).

Our approach involves a strategy to adapt δ dynamically during the search,
starting from $\delta = 1$. Once a maximum number of non-improving iterations ii_{max}
has been spent on a certain value of δ, this value is increased by one. The rationale

behind this is that, if the search cannot improve anymore by relaxing one day, then we try relaxing two days at once, and so on. Moreover, the value of δ is reset to 1 when a new best solution is found. When LNS has performed ii_{max} iterations on $\delta = h$ (the maximum number of relaxable days) the search is restarted.

A neighboring solution is accepted if its cost is lower or equal to the one of the current best solution. This allows to escape plateaux in the fitness landscape.

4   Experimental Analysis

The CP model, the custom propagator, and the branching rules were all implemented using the GECODE framework [12]. Moreover, the LNS method has been implemented as a generic search meta-engine for GECODE.

To assess the performance of our hybrid approaches, we have generated a number of random instances based on the presented problem formulation. The behavior of our instance generator[3] is influenced by a number of parameters. Among these are the geographical area, the planning horizon in days, the number of daily activities, a caregiver correction rate (a multiplier that can be used to increase or reduce the number of needed caregivers, which is computed heuristically), the types and probabilities of shifts for caregivers, the probability of a caregiver/activity incompatibility, the probability of an activity to need multiple caregivers, and the parameter k for the Poisson distribution from which the number of caregivers for an activity is drawn.

Overall, our benchmark set is composed of 18 families of 30 instances each (totaling 540 instances), differentiated by planning horizon, number of activities, and caregiver correction rate (c). In these instances, whose activities are located in a 40Km $\times$ 40Km rectangular area centered on Udine (Italy), the number of maximum consecutive days is always $h - 1$, while the other parameters have their default values. According to the best practices, we have tuned the parameters of our LNS approach ($ii_{max} = 60$ and $t_{var} = 10$ms) by running, through JSON2RUN [17], an $F\text{-}Race(RSD)$ [18] over a subset of 360 instances, and used the remaining 180 for validation against the pure CP approach. Both the training and validation instances are distributed together with the instance generator.

Table 2 summarizes, for each family of the validation instances, the aggregated mean cost (f) in €, number of unassigned activities (f_u), distance in Kilometers (f_d), hours of overtime (f_o), and hours of slack time (f_s) attained, respectively, by pure CP and our hybrid approach in 10 minutes. The last column shows the cost improvement attained when using LNS instead of pure CP.

From the results, it is easy to see that the LNS approach outperforms pure CP on all except one family of instances, mostly because of its superiority in dealing with unassigned activities (the most weighted cost component) and overtime work. On the other hand, CP seems to be better at reducing the total traveling distance. This is likely due to our branching strategy, which is able to reconsider

---

[3] Available at https://bitbucket.org/tunnuz/homecare-instance-generator

Table 2. Comparison between performance of CP (branch & bound) and LNS

Group			CP					LNS					Imp.
h	n_d	c	f	f_u	f_d	f_o	f_s	f	f_u	f_d	f_o	f_s	
3	20	0.6	2247.46	11	399.40	0.32	0.50	**961.37**	2	419.30	0.10	0.30	57.2%
3	20	0.7	1210.38	0	478.20	0.15	0.67	**1059.02**	0	367.80	0.04	0.72	12.5%
3	20	0.8	1358.65	4	345.17	0.14	0.53	**975.08**	0	341.20	0.06	0.64	28.2%
3	30	0.6	2621.19	9	512.43	0.41	0.90	**1220.18**	0	773.53	0.13	0.42	53.5%
3	30	0.7	2908.59	12	538.43	0.28	1.00	**1480.50**	0	591.67	0.27	0.72	49.1%
3	30	0.8	2255.51	9	537.23	0.13	0.78	**1265.01**	0	538.90	0.08	0.73	43.9%
3	40	0.6	4502.60	26	780.67	0.53	0.73	**1851.14**	0	807.93	0.18	0.99	58.9%
3	40	0.7	**2710.03**	4	833.37	0.58	1.05	2968.33	7	867.03	0.45	1.10	-9.5%
3	40	0.8	3330.83	7	785.37	0.52	1.48	**2405.65**	3	948.50	0.33	0.97	27.8%
6	20	0.6	3846.61	19	696.23	0.35	1.03	**2017.17**	0	787.97	0.10	1.27	47.6%
6	20	0.7	2713.60	5	824.00	0.32	1.21	**1605.11**	0	734.57	0.11	0.87	40.8%
6	20	0.8	3088.61	6	836.23	0.42	1.39	**1589.25**	0	725.83	0.08	0.90	48.5%
6	30	0.6	5862.63	27	1114.03	0.76	1.49	**2188.90**	0	1147.67	0.31	0.90	62.7%
6	30	0.7	4371.72	5	1137.47	0.85	2.12	**2922.87**	0	1060.97	0.40	1.65	33.1%
6	30	0.8	4000.81	5	1160.90	0.38	2.18	**2595.11**	1	1154.57	0.33	1.18	35.1%
6	40	0.6	7646.02	36	1557.80	1.02	1.73	**4096.22**	0	1542.47	0.57	2.25	46.4%
6	40	0.7	6721.86	17	1562.07	0.90	2.86	**4827.88**	0	1693.20	0.79	2.65	28.2%
6	40	0.8	6623.41	11	1634.90	0.93	3.30	**5486.30**	7	1540.33	0.73	2.81	17.2%

the caregivers' routes, exploiting the fact that it does not have a bound on the time, unlike LNS repair step.

5 Conclusions and Future Work

In this paper we tackle the problem of multi-day scheduling of homecare activities by means of two CP-based solution methods. The contributions of the paper are the following ones. First, we propose an effective routing-based CP-model for the problem, including custom look-ahead constraints, and a dedicated branching strategy. Second, we devise an adaptive LNS method, which considerably improves the results found by pure CP tree-search.

Among the alternatives we want to explore, are different relaxation methods for the LNS destroy step, and more refined branching strategies either for the determination of caregivers' days off or for a finer-grained construction of the routes. Moreover, we plan to test future refinements of this approach over real-world instances as soon as they will be provided by our industrial partner.

Acknowledgments. This work has been partially funded by Google Inc. under the Google Focused Grant Program on "Mathematical Optimization and Combinatorial Optimization in Europe". We thank EasyStaff, for providing us with the real-world problem specification.

References

1. Begur, S.V., Miller, D.M., Weaver, J.R.: An integrated spatial DSS for scheduling and routing home-health-care nurses. Interfaces 27(4), 35–48 (1997)
2. Cheng, E., Rich, J.L.: A home health care routing and scheduling problem. Technical Report CAAM TR98–04, Rice University (1998)
3. Rasmussen, M.S., Justesen, T., Dohn, A., Larsen, J.: The home care crew scheduling problem: Preference-based visit clustering and temporal dependencies. European Journal of Operational Research 219(3), 598–610 (2012)
4. Eveborn, P., Flisberg, P., Rönnqvist, M.: Laps care—an operational system for staff planning of home care. European Journal of Operational Research 171(3), 962–976 (2006)
5. Bertels, S., Fahle, T.: A hybrid setup for a hybrid scenario: Combining heuristics for the home health care problem. Computers and Operations Research 33(10), 2866–2890 (2006)
6. Rendl, A., Prandtstetter, M., Hiermann, G., Puchinger, J., Raidl, G.: Hybrid heuristics for multimodal homecare scheduling. In: Beldiceanu, N., Jussien, N., Pinson, É. (eds.) CPAIOR 2012. LNCS, vol. 7298, pp. 339–355. Springer, Heidelberg (2012)
7. Trautsamwieser, A., Hirsch, P.: Optimization of daily scheduling for home health care services. Journal of Applied Operational Research 3, 124–136 (2011)
8. Allaoua, H., Borne, S.: A matheuristic approach for solving a home health care problem. Electronic Notes in Discrete Mathematics 41, 471–478 (2013)
9. Steeg, J., Schröder, M.: A hybrid approach to solve the periodic home health care problem. In: Kalcsics, J., Nickel, S. (eds.) Operations Research Proceedings 2007. Operations Research Proceedings, vol. 2007, pp. 297–302. Springer, Heidelberg (2008)
10. Nickel, S., Schröder, M., Steeg, J.: Mid-term and short-term planning support for home health care services. European Journal of Operational Research 219(3), 574–587 (2012)
11. Kilby, P., Shaw, P.: Handbook of Constraint Programming: Vehicle Routing. Elsevier, New York (2006)
12. Schulte, C., Tack, G., Lagerkvist, M.Z.: Modeling and Programming with Gecode (2014), http://www.gecode.org/doc-latest/MPG.pdf
13. Shaw, P.: Using constraint programming and local search methods to solve vehicle routing problems. In: Maher, M.J., Puget, J.-F. (eds.) CP 1998. LNCS, vol. 1520, pp. 417–431. Springer, Heidelberg (1998)
14. Pisinger, D., Ropke, S.: Large neighborhood search. In: Handbook of Metaheuristics, pp. 399–419. Springer (2010)
15. Bent, R., Van Hentenryck, P.: A two-stage hybrid local search for the vehicle routing problem with time windows. Transportation Science 38(4), 515–530 (2004)
16. Rousseau, L.M., Gendreau, M., Pesant, G.: Using constraint-based operators to solve the vehicle routing problem with time windows. Journal of Heuristics 8(1), 43–58 (2002)
17. Urli, T.: json2run: a tool for experiment design & analysis. CoRR abs/1305.1112 (2013)
18. Birattari, M., Yuan, Z., Balaprakash, P., Stützle, T.: F-Race and iterated F-race: An overview. Springer, Berlin (2010)

A Hybrid Metaheuristic to Solve the Resource Allocation Problem in Bike Sharing Systems

Patrick Vogel, Bruno A. Neumann Saavedra, and Dirk C. Mattfeld

Decision Support Group, Technische Universität Braunschweig,
Mühlenpfordtstraße 23, 38106 Braunschweig, Germany
{p.vogel,b.neumann,d.mattfeld}@tu-braunschweig.de

Abstract. Bike sharing systems have recently enabled sustainable means of shared mobility through automated rental stations. Spatio-temporal variation of bike rentals, however, leads to imbalances in the distribution of bikes causing full or empty stations.

The resource allocation problem tackles imbalances at a tactical planning level by means of bike allocation and relocation. We propose a MIP formulation of an extended dynamic service network design model. The objective is to determine optimal fill levels at stations while minimizing the expected costs of relocation for the typical bike demand. The MIP formulation is hard to solve due to a large number of binary variables for relocations (stations times stations times periods).

Thus, we present a hybrid metaheuristic integrating a large neighborhood search with exact solution methods provided by a solver. The large neighborhood search iteratively improves the solution with the help of limiting and controlling possible relocation regimes by a fix-and-optimize strategy, i.e. a small subset of "free" binary relocation variables. The majority of remaining binary variables are tentatively fixed to zero leading to a fast solvable truncated MIP of the resource allocation problem. Therefore, a commercial solver can provide a local optimal value based on the defined neighborhood, in a reasonable time. Results obtained indicate that the hybrid metaheuristic outperforms CPLEX for data from Vienna's bike sharing system "Citybike Wien".

Keywords: Hybrid metaheuristic, large neighborhood search, fix-and-optimize, mixed-integer programming, resource allocation, bike sharing.

1 Logistical Challenges in Bike Sharing Systems

Bikes are gaining more and more attention as an alternative and sustainable mode of transportation in urban areas. In recent years, innovative bike sharing systems (BSS) have been implemented in about 400 cities in Europe [1]. BSS provide bikes at unattended bike stations city-wide [2]. Prominent BSS grant that the first 30 minutes of trips are free. The operation of BSS will likely need financial support since money often cannot be directly earned with BSS [1].

Information systems enable fully automated rental and return operations, inducing one-way trips and potentially short rental times. Due to the mobility needs of users, spatio-temporal demand variation occurs leading to imbalances in bike distribution.

M.J. Blesa, C. Blum, and S. Voß (Eds.): HM 2014, LNCS 8457, pp. 16–29, 2014.

Depending on the location, stations may run full or empty or may have cyclic demand variation in the course of day. For instance, stations that tend to run empty are located on hills whereas stations in a residential area run empty in the morning and full in the evening due to commutes.

Thus, BSS operators face challenging logistical tasks regarding the reliability of service. For instance, a tendering for the Arlington BSS requests that "stations shall not be full of bicycles for more than 60 minutes during the hours of 8am - 6pm and 180 minutes during the hours of 6pm - 8am" [3]. Operators have to ensure bike availability for rentals as well as a sufficient number of free bike racks for returns by relocating bikes from full to empty bike stations. However, relocation of bikes with the help of a service fleet results in significant costs affecting the viability of BSS [4].

On tactical level, the required relocation demand satisfying typical demand in terms of bike flows is determined. The resource allocation problem (RAP) deals with the adequate distribution of bikes by means of allocation whereas relocation operations are anticipated. Therefore, fill levels of bikes at stations need to be determined in order to compensate typical bike demand variation in the course of day. Anticipation of relocation demand occurs by determining relocation operations abstracting from relocation tours on the operational level. A relocation operation consist of the pickup and return station as well as the time-period and number of relocated bikes. Decisions on allocation and relocation are interdependent, since reasonable fill levels of bikes may reduce relocation efforts, whereas high relocation efforts may compensate insufficient bike fill levels. Typical bike flows serve as input occurring due to recurring mobility patterns in the course of day. Thus, the relocation operations are computed once, e.g. for the typical working day in the summer season. The determined relocation operations then have to be implemented in relocation tours and possibly adjusted to the demand variation on the operational level.

In this paper, we propose a MIP formulation of the RAP. The presented mathematical model determines the optimal fill level at stations by minimizing the expected costs of relocation while ensuring a given service level. Since the MIP is hard to solve due to a high number of binary variables modeling relocation operations, we propose a hybrid metaheuristic (HM). The HM consists of the definition and further exploration of a neighborhood based on large neighborhood search (LNS) guided by a fix-and-optimize strategy and exact solution methods provided by a solver.

The remainder of this paper is organized as follows. A literature overview on the optimization of BSS is given in Section 2. We present the MIP formulation of the RAP in Section 3. The hybrid optimization approach is subject to Section 4. The proposed methodology is exemplified with the help of a case study including trip data from Vienna's BSS "Citybike Wien" (Section 5). Future work is subject of Section 6.

2   Related Literature

The optimization of BSS has become very active over the last years. The majority of papers focuses solely on the operational planning of relocation tours. Articles on the tactical planning of bike allocation as well as the integrated planning of allocation and relocation are rather scarce. The following literature review presents operational, tactical and integrated planning approaches.

In many articles, the optimization of relocation tours is based on the one commodity pickup and delivery problem (PDP) and the swapping problem (SP). The PDP deals with a fleet of vehicles transporting a commodity from pickup to delivery stations. The SP deals with multiple commodities and a station serves as both a pickup and delivery station. Modeling of relocations can be further classified into static or dynamic. In the static case, relocations are usually realized at the time of the day with the lowest overall demand, e.g. early morning hours. In the dynamic case, demand variation and several decision points over time are considered.

Benchimol et al. [5] combine PDP and SP and present a static relocation model as well as solution methods. Raviv et al. [6] study the static relocation problem minimizing user dissatisfaction by means of penalty costs and operating costs for relocation. Rainer-Harbach et al. [7] introduce a combined variable neighborhood search (VNS) and greedy heuristic for a maximum flow approach and linear program to determine the routes and number of relocated bikes for the static relocation problem. Raidl et al. [8] improve the aforementioned approach by efficiently determining optimal loading operations. Also addressing the static case, Di Gaspero et al. [9] present a HM combining constraint-based programming (CP) and ant colony optimization. The objective is to minimize the travel time for relocation tours and difference between actual and target fill level at stations. Di Gaspero et al. [10] extend the CP approach and incorporate LNS to speed up to the branching strategy inherent in the CP. Ricker et al. [11] use a simulation-based approach to obtain the cost-efficient number of relocation operations in the course of day considering weighted sums of transportation costs and costs for unserved customers. Contardo et al. [12] present an arc-flow optimization model for the dynamic routing of relocation vehicles minimizing "lost demand". Customers who cannot rent or return rented vehicles at empty or full stations cause lost demand. Caggiani and Ottomanelli [13] propose a decision support system for the dynamic relocation problem. Here, a neural network is used to forecast rentals and returns at stations. Dell'Amico et al. [14] present MIP formulations addressing the dynamic relocation problem. Due to an exponential number of constraints, a branch-and-cut algorithm for solving is implemented.

Regarding the tactical planning, George and Xia [15] apply closed queuing network to model the underlying system. A profit maximizing optimization is used to determine the optimal fleet size and allocation of rental vehicles. Raviv and Kolka [16] also pursue a queuing model approach. Based on a user dissatisfaction function, the optimal fill level at one bike station is determined.

The work presented above focusses either the tactical or the operational planning level. Integrated approaches are scarce. Especially for tactical planning tasks, anticipation of operational decisions is crucial for the viability of BSS, since costly relocation can be alleviated by fill levels compensating expected demand variation. To the best of the authors' knowledge, the following approaches integrating allocation and relocation exist:

- Correia and Antunes [17] present multi-periodic MIP formulations to maximize the profit of a car sharing system. The revenue of trips, costs of depot and vehicle maintenance as well as costs of vehicle relocation are considered. Optimization determines the number and the location of stations as well as the number of vehicles at every station in each period of daily operation. They consider static relocation at the end of the day allowing relocation of vehicles between stations to reset the initial fill level. A simulation model validates the validity of the MIP [18].
- Sayarshad et al. [19] introduce a dynamic LP formulation to maximize profit in BSS. The objective function subtracts relocation, maintenance, capital investment and holding costs of bikes as well as penalty costs for lost demand from revenue generated by trips. Bikes can be relocated in every period of daily operation.
- Cepolina and Farina [20] determine the fleet size and vehicle allocation for a system with small electronic vehicles. Costs for user waiting times and system operation (vehicle purchasing and running costs) are minimized. The authors state, that flexible users perform relocation operations under the supervision of the system provider.
- Schuijbroek et al. [21] minimize the costs of relocation tours and incorporate service level requirements at stations. They consider the static case in which no user demand occurs whereas the service level is precalculated for each station. A cluster-first route-second heuristic is proposed to solve the problem.

In sum, none of the recent approaches sufficiently covers both the dynamic interaction of bikes in BSS and minimizing costs from relocation operations. Thus, we propose an adequate integrated approach in the subsequent section.

3   The Resource Allocation Problem

The following MIP optimization model tackles the described tactical RAP. Decisions at this planning level are somehow abstracted and should apply to a wide-range of data and system parameters. We follow the work of Crainic [22] on tactical service network design in freight transportation. Decisions aim at the optimal allocation and utilization of resources to satisfy customer service and economic goals. The objective is to determine offered transportation services between nodes within the network and required capacities on the links at the lowest costs. In the case of BSS, the service operator transports bikes in capacitated trucks from full to empty stations to maintain service levels and associated fill levels. Thus, fixed transportation costs for providing relocation operations and variable costs for the particular handling of transported bikes occur. Input of the RAP are typical bike flows in the form of time-dependent origin-destination matrices reflecting daily mobility patterns. For more information on the typical system behavior and a LP formulation of the RAP see Vogel et al. [23].

In the following MIP formulation, we minimize the total expected costs of system operation occurring from relocation while satisfying a minimal service level. Output of the optimization are the total expected costs for relocation and potentially unsatisfied demand. Moreover, the optimization yields relocation operations and fill levels at each station in the course of day.

Input of the optimization covers the configuration of the BSS and demand obtained from the BSS information system. The system's configuration comprises a set of rental stations N each providing s_i bike racks and a total number of b bikes in the system. Considering T periods in the course of day, the number of available bikes at a station i and in period t is B_{it}. The number of available bikes depends on the bike demand and relocated bikes. Bike demand is given by the bike flows $f_{ij,t}$ between stations i and j in a period t. Relocation operations $RO_{ij,t}$ are implemented compensating missing bikes or bike racks by relocating a certain number of bikes $R_{ij,t}$. The objective is to minimize the total costs of relocation ensuring safety buffers of bikes sb_{it} and bike racks sbr_{it}. If safety buffers are violated, additional costs for missing bikes MB_{it} and missing bike racks MBR_{it} occur. The resource allocation model reads as follows:

Sets

- $N = \{1, \ldots, n\}$: set of stations
- $T = \{0, \ldots, t_{max}\}$: set of periods, e.g., hours of the day. For resetting the number of allocated bikes at the end of the day, t_{max} includes the first period of the next day

Decision variables

- $B_{i,t} \in \mathbb{R}$: number of bikes at station i in time period t
- $R_{ij,t} \in \mathbb{R}$: relocated bikes between stations i and j in time period t
- $RO_{ij,t} \in \{0,1\}$: relocation operation between stations i and j in time period t
- $MB_{i,t} \in \mathbb{R}$: number of missing bikes at station i in time period t
- $MBR_{i,t} \in \mathbb{R}$: number of missing bike racks at station i in time period t

Parameters

- s_i : size of stations in terms of bike racks at station i
- b : total number of bikes in the system
- $f_{ij,t}$: bike flow between stations i and j in time period t
- ch_t : average handling costs of one relocated bike in time period t
- ct_{ij} : average transportation costs of a relocation operation between stations i and j
- cm : penalization cost per missing bike and per missing rack bike
- $pb_{i,t} \in [0,1]$: proportion of returned bikes that are available for rentals at station i in time period t
- $pbr_{i,t} \in [0,1]$: proportion of rented bikes that are available for returns at station i in time period t
- l : lot size (capacity) for relocation operations
- $sb_{i,t}$: bike safety buffer at station i in time period t
- $sbr_{i,t}$: bike rack safety buffer at station i in time period t

With this notation the optimization model reads:

$$\text{Minimize} \quad \sum_{t=0}^{t_{max}} \sum_{i=1}^{n} \sum_{j=1}^{n} \left(ch_t \cdot R_{ij,t} + ct_{ij} \cdot RO_{ij,t}\right) + cm \cdot \left(MB_{i,t} + MBR_{i,t}\right) \quad (1)$$

subject to

$$B_{i,t+1} = B_{i,t} + \sum_{j=1}^{n}\left(f_{ji,t} - f_{ij,t} + R_{ji,t} - R_{ij,t}\right) \quad \forall i \in N, t \in T \backslash t_{max} \quad (2)$$

$$B_{i,t} - \sum_{j=1}^{n} f_{ij,t} + pb_{i,t} \sum_{j=1}^{n} f_{ji,t} - \sum_{j=1}^{n} R_{ij,t} + MB_{i,t} \geq sb_{i,t} \; \forall i \in N, t \in T \quad (3)$$

$$s_i - B_{i,t} - \sum_{j=1}^{n} f_{ji,t} + pbr_{i,t} \sum_{j=1}^{n} f_{ij,t} - \sum_{j=1}^{n} R_{ji,t} + MBR_{i,t} \geq sbr_{i,t} \; \forall i \in N, t \in T \quad (4)$$

$$l \cdot RO_{ij,t} \geq R_{ij,t} \; \forall i,j \in N, t \in T \quad (5)$$

$$R_{ij,0} = 0 \; \forall i,j \in N \quad (6)$$

$$B_{i,0} = B_{i,t_{max}} \; \forall i \in N \quad (7)$$

$$\sum_{i=1}^{n} B_{i,t} = b \; \forall t \in T \quad (8)$$

$$B_{i,t}, R_{ij,t}, RO_{ij,t}, MB_{i,t}, MBR_{i,t} \geq 0 \; \forall i,j \in N, t \in T \quad (9)$$

The objective function (1) minimizes the relocation costs comprising handling costs for each individual bike and one-time cost for the particular transport. In addition, costs for violating service levels by means of slack variables for missing bikes or missing bike racks occur. The slack variables guarantee feasibility of the model even if demand exceeds capacity. Equation (2) ensures flow conservation, i.e., the number of bikes in a station in the next period depend on the number of bikes in the current period plus returned bikes by users and relocation and minus rented bikes by users and relocation. It is assumed, that relocation operations last one time-period, e.g. one hour. However, relocation time can be adjusted. Constrains (3) and (4) are related to the service level offered by the system. Constraint (3) ensures a minimal safety buffer of bikes at stations, i.e. the number of allocated bikes minus customer rentals plus a certain proportion of returns $pb_{i,t}$ minus relocation pickups plus potentially missing bikes. We explicitly model the proportion of returns due to temporal aggregation of trips since information on the sequence of rentals and returns is lost, e.g. if trips are hourly aggregated to bike flows. By setting $pb_{i,t}$ to 1, rentals and returns are instantly interchanged within a period. By setting $pb_{i,t}$ to 0, rentals and returns are considered separately. Constraint (4) similarly ensures safety buffer for bike racks. The relocation demand is aggregated to operations by modeling lots. Thus, equation (5) aggregates relocations by means of a binary variable and the capacity is given by the lot size l. Relocation is prohibited in the first period (6). The typical demand reflects recurring mobility patterns. Equation (7) ensures that fill levels at the end of the day match the initial fill levels due to recurring demand patterns. However, the end fill levels can be adjusted, e.g. transition from working day to weekend. All existing bikes need to be allocated at the stations (8). Decision variables must be non-negative (9).

4   The Hybrid Metaheuristic

The proposed model can be seen as specialization of the service network design [22] and therefore belongs to the class of NP-hard problems. Especially real-world instances are computational challenging since typical system sizes range from 50 to more than 1000 bike stations. Thus, excessive computation time is required to find the optimal solution since the model stands out due to a large number of binary variables for relocation operations (stations x stations x periods). That is why, we propose a HM integrating LNS guided by a fix-and-optimize strategy and exact solution techniques provided by a solver to tackle the RAP. First, we motivate the use of a HM and sketch solution approaches related to our HM. Second, the basic idea of our HM is outlined followed by the algorithmic description. Last, the used neighborhood operators of LNS are presented.

Blum and Roli [24] state that metaheuristics show good performance for combinatorial problems that are computationally intractable for large instances. However, hybridization of metaheuristics leads even more powerful solution methods. Hybridization occurs by e.g. integrating metaheuristics with classical operation research methods [25]. Hence, they offer flexibility to combine different algorithm components with other mathematical programming techniques such as branch and bound and therefore provide good strategies for solving large-scale problems. Blum et al. [26] give a recent survey and classification of HMs.

We follow the HM approach by Shaw [27] based on LNS and using a constraint-based tree search originally applied to the Vehicle Routing Problem. Di Gaspero et al. [10] also follow Shaw tackling the operational relocation problem in BSS. CP-based LNS exploits CP algorithms such as branch and bound to explore a large neighborhood. Basically, each iteration of the local search is the solution of a set of the search space [28]. Basic idea of the approach is to improve an initial solution by removing and re-inserting customer visits, i.e. fixing variables to one or zero, by CP.

The basic idea of our hybridization approach (cf. Fig. 1) is to guide LNS by a fix-and-optimize strategy to iteratively limit and control the optimization problem to a tractable set of "free", i.e. $\in \{0,1\}$, binary variables. Our approach shows similarities to Pisinger and Ropke's [29] adaptive LNS using fix-and-optimize operations as destroy and repair operators. In the case of the RAP, LNS generates a neighborhood by freeing a tractable set of binary relocation variables and fixing the majority of binary variables to zero leading to a truncated, fast solvable version of the RAP. For the truncated RAP, the solver decides which relocation operations provide an optimal solution, i.e. assigning either zero or one to the free binary variables. In particular, the algorithm iteratively improves a solution by exploring in each iteration p a neighborhood K^p of a relocation solution based on LNS. The solver returns the new local optimal solution S^p for the truncated RAP that is further adapted by LNS. An initial solution S^0 of the RAP with the full set of free binary variables is obtained by a solver with restricted running-time.

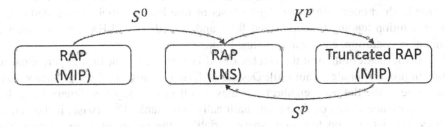

Fig. 1. Proposed hybridization approach

We refrain from the common fix-and-optimize strategy fixing binary variables to either one or zero. In the case of the RAP, this approach counteracts the dependencies of allocation and relocation due to the spatio-temporal demand characteristics of the bike station network. We assume for one moment that all binary variables are fixed to either one or zero. In addition, we presume that the system infrastructure can provide bikes and bike racks to satisfy all the service level. If many variables are fixed to zero, slack variables guarantee to always find a feasible solution. However, assigned slack variables result in a low quality solution. In contrast, if many variables are fixed to one, all the demand will be likely satisfied, but results in a low quality solution due to the high fixed costs. Thus, we prove free binary variables to the solver. Then the solver decides on setting free binary variables to either one or zero.

```
Input: RAP Problem
Output: Best solution found
p ← 0
Sᵖ ← InitialSolution()
While termination condition not met do
    p ← p + 1
    Operators to explore the neighborhood Kᵖ by selecting v
    of the fixed ROᵢⱼ,ₜ variables:
    Kᵢᵖ ← INSERT(Sᵖ⁻¹)
    Kₑᵖ ← EXCHANGE(Sᵖ⁻¹)
    Kₛᵖ ← SHIFT(Sᵖ⁻¹)
    Kᵖ = Sᵖ⁻¹ ∪ Kᵢᵖ ∪ Kₑᵖ ∪ Kₛᵖ
    Let RAP|ₖᵖ be the problem in which variables ROᵢⱼ,ₜ(Kᵖ)
    are free.
    Sᵖ ← SOLVE(Kᵖ);
End while
```

The proposed HM starts by generating an initial solution obtained by CPLEX. CPLEX will not likely find an optimal solution because of the high number of binary variables. Thus, the solver runs only for a short time to find a feasible solution S^0. To improve the solution, we define in each iteration p a neighborhood K^p by means of adding a set of free binary variables to the former solution exploiting the characteristics of relocation operations in practice. The neighborhood operators are described below in detail. Then we solve the truncated RAP with CPLEX returning the optimal decision in short computing time. Again, a set of free binary variables is added to the former solution leading to a truncated RAP and solved by CPLEX until we meet a termination condition, e.g. a running-time limit.

For LNS, the definition of the neighborhood is very important for the performance of this improvement algorithm [30]. Despite of fixing a substantial set of binary variables to zero, the defined neighborhood may result in to many free binary variables and the truncated RAP could be computationally intractable. Otherwise, if the neighborhood consists of too few free binary variables, the possibilities to improve the solution are low. Therefore, the decision on the number of free binary variables is not trivial. However, a suitable number can be estimated based on previous computational tests.

Regarding the definition of the neighborhood, it is recommended to use filtering techniques or criteria based on the characteristics of the model to explore the search space efficiently [31]. We follow Ahuja [30] and define the neighborhood operators by exploiting characteristics of RAP. In particular, promising relocation operations, i.e. free binary variables, are added. The *EXCHANGE* and *SHIFT* neighborhood operators exploit spatio-temporal variations of relocations of the feasible solution. In addition, *INSERT* provides randomly selected relocation operations to explore the feasible space with more diversity. We propose that relocation selected to define a neighborhood are preferably between near stations since the transportation costs are lower. Thus, a roulette wheel selection is implemented giving close stations a higher probability. Please note, that a specific destroy operator is not implemented since CPLEX decides on assigning binary variables. In particular, we present the following three neighborhood operators (cf. Fig. 2):

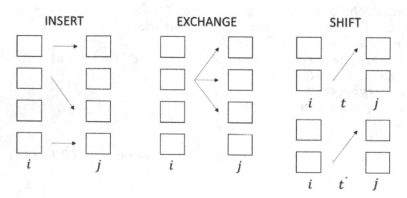

Fig. 2. Neighborhood operators

- *INSERT*: additional relocation operations are added. The source station i is chosen randomly whereas the sink station j is selected depending on the distance based on the roulette wheel. This allows to explore zones of the search space without a direct relation to the solution, providing diversity.
- *EXCHANGE*: stations that provide bikes to other stations in some periods can be regarded as surplus stations. Thus, additional relocation operations from these source stations to other stations are freed. Again, a specific surplus station is randomly chosen from all surplus stations and the sink station is selected depending on the distance based on the roulette wheel.
- *SHIFT*: a source station for an existing relocation operations is randomly selected and shifted to other time periods. Here, the intervals of two adjacent time periods have a higher probability whereas other periods have a lower probability.

Finally, the solver is executed. Thus, a new feasible solution s is obtained. In the worst case, the previous feasible solution may be obtained.

5   Computational Study

A case study is used to evaluate the performance of the HM. Input data are obtained based on the analysis of Vienna's BBS "Citybike Wien" (CBW). CBW provided data comprising the years 2008 and 2009. In that time, CBW consisted of n = 59 stations placed around the city with a given number of bike racks and a total of $b = 627$ bikes (~50% fill level). Time is discretized in terms of $t_{max} = 24$ (hour) periods, reflecting typical aggregation from the field of traffic analysis. In the case of Vienna, 92 % of the trips are shorter than 60 minutes. Moreover, almost 70% of the trips end within the same hour, e.g., a trip that starts at 4:xx will end at 4:xx. We assume that relocation operations take one hour on average. Handling costs depend on the time of day and are set to $ch_{day} = 4$ Euro (in effect for periods 9 to 20), while night time handling costs are more expensive ($ch_{night} = 7$ Euro). Transportation costs are assumed independent of the time of the day and amount to $ct_{ij} = 0.5$ Euro per kilometer. Bike and bike rack safety buffers are set to 0 avoiding the use of slack variables. Data mining tools are used to estimate the bike flows in 24 periods of the typical working day [23]. In particular, we use instances with low, medium and high demand measured in the number of trips per bike per day. The average number of trips per bike and day in the summer season is 2.5 (1569 trips in total) and regarded as the low demand instance. Furthermore, the medium demand instance has 4 and the high demand instance has 6 trips per bike and day. We evaluate scenarios in which all demand is satisfied avoiding the use of slack variables since penalization cost can be a significant proportion of the total cost and therefore distort the computational analysis of the algorithm. The MIP was implemented in IBM ILOG OPL. The HM was implemented in Java using the ILOG Concert Technology to access CPLEX. Both approaches use CPLEX 12.5 on an Intel® Core™2 Duo CPU at 3.17GHz processor with 6.00 GB RAM.

Table 1 shows the obtained results for the different demand instances and operator combinations (INSERT=I, EXCHANGE=E, SHIFT=S). Initial tests showed, that CPLEX provides an optimal solution for $v = 500$ "free" binary variables within seconds. The mix of operators is tested in five runs each. We provide the average number of p iterations that the HM realizes in the runs, the best found solution and the average solution over the five runs. Both CPLEX and the HM were running for 20 minutes. The CPLEX solution is used as reference for comparison of CPLEX and HM. In particular, we measure the improvement of the HM according to best obtained solution by $\frac{MIP\ BEST - HM\ BEST}{HM\ BEST} \cdot 100\%$ and the improvement of the average solution analogously.

The proposed HM seems stable providing good quality solutions since the HM yields lower costs in many of the tested runs of our case study. The three neighborhood operators are tested independently to determinate their effect separately. The results show that INSERT and EXCHANGE yield better solutions than only using CPLEX to solve the full RAP. Spatial variation of relocations therefore effect the solution quality. The SHIFT operator produces low-quality solutions indicating that the effect of only temporal variation on relocations is small. The combination of spatial and temporal operators INSERT, EXCHANGE and SHIFT as well as the combination of both spatial operators INSERT and EXCHANGE yield good results. In addition, the best and average solutions of the HM, except for the SHIFT operator, are always better than the CPLEX solution. However, the proposed HM only obtains improvements up to 0.5%. Nevertheless, we are quite confident that the improvement increases for bigger instances.

Table 1. Computational results of the HM operators

Demand	Operator	v	p (AVG)	MIP CPLEX Best	GAP	Hybrid Metaheuristic Best	AVG	Improv. Best	Improv. AVG
Low				495.899	1.96%				
	I	500	665.8			495.539	495.959	0.07%	-0.01%
	E	500	226			495.143	495.302	0.15%	0.12%
	S	500	1445.6			500.477	500.477	-0.91%	-0.91%
	I,E,S	100,350,50	182.8			495.255	495.411	0.13%	0.10%
	I,E	125,375	185			495.195	495.462	0.14%	0.09%
Medium				773.163	1.53%				
	I	500	655			769.543	769.861	0.47%	0.43%
	E	500	204.8			769.332	769.441	0.50%	0.48%
	S	500	1260.8			775.125	775.125	-0.25%	-0.25%
	I,E,S	100,350,50	174			769.117	769.600	0.53%	0.46%
	I,E	125,375	197			769.390	769.811	0.49%	0.44%
High				1120.611	1.01%				
	I	500	829.2			1118.866	1119.617	0.16%	0.09%
	E	500	204			1119.167	1119.581	0.13%	0.09%
	S	500	1352.8			1125.715	1125.715	-0.45%	-0.45%
	I,E,S	100,350,50	179.6			1119.132	1119.550	0.13%	0.09%
	I,E	125,375	170			1118.779	1119.405	0.16%	0.11%

Regarding the different demand instances, there is no superior operator combination. In the low demand case, the EXCHANGE operator yields the best solution. For medium demand the INSERT, EXCHANGE and SHIFT and for high demand the INSERT and EXCHANGE combinations are best. In order to give insights on the solution strategy, the run of the HM with INSERT, EXCHANGE and SHIFT is exemplified for the medium demand instance (cf. Fig. 3).

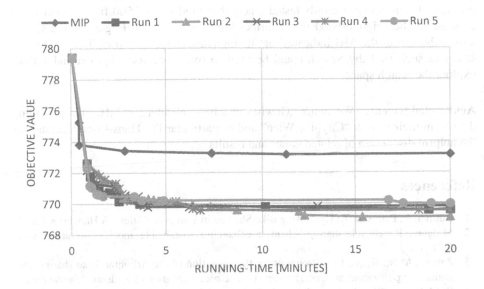

Fig. 3. The run of the CPLEX and MH (I,E,S) solution for the medium demand instance

CPLEX quickly improves the solution after a few seconds. However, afterwards it only marginally progresses in finding better solutions. The MH is slower in the beginning. It needs approximately one minute to improve the initial solution, but obtained solutions outperform CPLEX in all runs. In the first five minutes, several better solutions are found that seem to be rather diverse. After that, solutions become similar. Nevertheless, they fade towards different local optimal solutions.

These results are promising. Still we observe that it is increasingly difficult to find a neighborhood that improves a good-quality solution. The implementation of some techniques to penalize certain areas of the feasible space that does not generate a better solution can be an alternative to improve the performance of our algorithm.

6   Conclusions and Outlook

In this paper we have proposed a MIP formulation of the resource allocation problem in BSS. Since the formulation is computational challenging, a HM was introduced to the solve the MIP. Experiments for a small real-world BSS show that our approach outperforms CPLEX. Although CPLEX obtains better solutions after a few seconds, the MH adds more diversity and thus finds better solutions in the long run. The combination of operators yielding a good solution depends on the instances and one superior combination cannot be determined.

It is up to future research to assess the performance of the HM for bigger instances. In particular, a large number of bike stations and bike flows might challenge the HM. Thus, the performance of our MH with the different operator combinations has to be assessed. Furthermore, we only tested a neighborhood size of 500 free variables. An increase of the neighborhood might further improve solutions. Regarding the search space, the run of the MH indicates that it stagnates in local optima. Thus, a control structure such as Tabu Search could be used to overcome local optima and further explore the search space.

Acknowledgements. We thank „Gewista Werbegesellschaft m.b.H." for providing data from their project "Citybike Wien" and in particular Dr. Hans-Erich Dechant for the helpful discussion of related issues and results.

References

1. Büttner, J., Petersen, T.: Optimising Bike Sharing in European Cities - A Handbook (2011)
2. Midgley, P.: Bicycle-sharing schemes: enhancing sustainable mobility in urban areas (2011)
3. Zahory, M.N.: Request for Proposals for the operation of the Arlington Bike-sharing Program, http://www.metrobike.net/index.php?s=file_download&id=18
4. DeMaio, P.: Bike-sharing: History, impacts, models of provision, and future. Journal of Public Transportation 12, 41–56 (2009)
5. Benchimol, M., Benchimol, P., Chappert, B., De La Taille, A., Laroche, F., Meunier, F., Robinet, L.: Balancing the stations of a self-service bike hire system. RAIRO-Operations Research 45, 37–61 (2011)
6. Raviv, T., Tzur, M., Forma, I.A.: Static repositioning in a bike-sharing system: models and solution approaches. EURO Journal on Transportation and Logistics, 1–43 (2012)
7. Rainer-Harbach, M., Papazek, P., Hu, B., Raidl, G.R.: Balancing bicycle sharing systems: A variable neighborhood search approach. In: Middendorf, M., Blum, C. (eds.) EvoCOP 2013. LNCS, vol. 7832, pp. 121–132. Springer, Heidelberg (2013)
8. Raidl, G.R., Hu, B., Rainer-Harbach, M., Papazek, P.: Balancing bicycle sharing systems: Improving a VNS by efficiently determining optimal loading operations. In: Blesa, M.J., Blum, C., Festa, P., Roli, A., Sampels, M. (eds.) HM 2013. LNCS, vol. 7919, pp. 130–143. Springer, Heidelberg (2013)
9. Di Gaspero, L., Rendl, A., Urli, T.: A hybrid ACO+CP for balancing bicycle sharing systems. In: Blesa, M.J., Blum, C., Festa, P., Roli, A., Sampels, M. (eds.) HM 2013. LNCS, vol. 7919, pp. 198–212. Springer, Heidelberg (2013)
10. Di Gaspero, L., Rendl, A., Urli, T.: Constraint-based approaches for balancing bike sharing systems. In: Schulte, C. (ed.) CP 2013. LNCS, vol. 8124, pp. 758–773. Springer, Heidelberg (2013)
11. Ricker, V., Meisel, S., Mattfeld, D.C.: Optimierung von stationsbasierten Bike-Sharing-Systemen. In: Proceedings of MKWI 2012, pp. 215–226 (2012)

12. Contardo, C., Morency, C., Rousseau, L.M.: Balancing a dynamic public bike-sharing system (2012)
13. Caggiani, L., Ottomanelli, M.: A modular soft computing based method for vehicles repositioning in bike-sharing systems. Procedia-Social and Behavioral Sciences 54, 675–684 (2012)
14. Dell'Amico, M., Hadjicostantinou, E., Iori, M., Novellani, S.: The bike sharing rebalancing problem: Mathematical formulations and benchmark instances. Omega (2013)
15. George, D.K., Xia, C.H.: Fleet-sizing and service availability for a vehicle rental system via closed queueing networks. European Journal of Operational Research 211, 198–207 (2011)
16. Raviv, T., Kolka, O.: Optimal inventory management of a bike-sharing station. IIE Transactions (2013)
17. Correia, G.H.A., Antunes, A.P.: Optimization approach to depot location and trip selection in one-way carsharing systems. Transportation Research Part E: Logistics and Transportation Review 48, 233–247 (2012)
18. Jorge, D., Correia, G.: Carsharing systems demand estimation and defined operations: a literature review. EJTIR 13, 201–220 (2013)
19. Sayarshad, H., Tavassoli, S., Zhao, F.: A multi-periodic optimization formulation for bike planning and bike utilization. Applied Mathematical Modelling 36, 4944–4951 (2012)
20. Cepolina, E.M., Farina, A.: A new shared vehicle system for urban areas. Transportation Research Part C: Emerging Technologies 21, 230–243 (2012)
21. Schuijbroek, J., Hampshire, R., van Hoeve, W.-J.: Inventory Rebalancing and Vehicle Routing in Bike Sharing Systems (2013)
22. Crainic, T.G.: Service network design in freight transportation. European Journal of Operational Research 122, 272–288 (2000)
23. Vogel, P., Ehmke, J.F., Mattfeld, D.C.: Decision support for tactical resource allocation in bike sharing systems (under review, 2014)
24. Blum, C., Roli, A.: Metaheuristics in combinatorial optimization: Overview and conceptual comparison. ACM Computing Surveys (CSUR) 35, 268–308 (2003)
25. Blum, C., Roli, A.: Hybrid metaheuristics: an introduction. In: Hybrid Metaheuristics. pp. 1–30. Springer (2008)
26. Blum, C., Puchinger, J., Raidl, G.R., Roli, A.: Hybrid metaheuristics in combinatorial optimization: A survey. Applied Soft Computing 11, 4135–4151 (2011)
27. Shaw, P.: Using constraint programming and local search methods to solve vehicle routing problems. In: Maher, M.J., Puget, J.-F. (eds.) CP 1998. LNCS, vol. 1520, pp. 417–431. Springer, Heidelberg (1998)
28. Pesant, G., Gendreau, M.: A view of local search in constraint programming. In: Freuder, E.C. (ed.) CP 1996. LNCS, vol. 1118, pp. 353–366. Springer, Heidelberg (1996)
29. Pisinger, D., Ropke, S.: A general heuristic for vehicle routing problems. Computers & Operations Research 34, 2403–2435 (2007)
30. Ahuja, R.K., Ergun, Ö., Orlin, J.B., Punnen, A.P.: A survey of very large-scale neighborhood search techniques. Discrete Applied Mathematics 123, 75–102 (2002)
31. Pisinger, D., Ropke, S.: Large neighborhood search. In: Handbook of Metaheuristics, pp. 399–419. Springer (2010)

Algorithm Comparison by Automatically Configurable Stochastic Local Search Frameworks: A Case Study Using Flow-Shop Scheduling Problems

Franco Mascia[1], Manuel López-Ibáñez[1], Jérémie Dubois-Lacoste[1],
Marie-Éléonore Marmion[2], and Thomas Stützle[1]

[1] IRIDIA, CoDE, Université Libre de Bruxelles, Brussels, Belgium
[2] LIFL, Université Lille 1, Inria Lille Nord-Europe
{fmascia,manuel.lopez-ibanez,jeremie.dubois-lacoste,stuetzle}@ulb.ac.be,
marie-eleonore.marmion@univ-lille1.fr

Abstract. The benefits of hybrid stochastic local search (SLS) methods, in comparison with more classical (non-hybrid) ones are often difficult to quantify, since one has to take into account not only the final results obtained but also the effort spent on finding the best configuration of the hybrid and of the classical SLS method. In this paper, we study this trade-off by means of tools for automatic algorithm design, and, in particular, we study the generation of hybrid SLS algorithms versus selecting one classical SLS method among several. In addition, we tune the parameters of the classical SLS method separately and compare the results with the ones obtained when selection and tuning are done at the same time. We carry out experiments on two variants of the permutation flowshop scheduling problem that consider the minimization of weighted sum of completion times (PFSP-WCT) and the minimization of weighted tardiness (PFSP-WT). Our results indicate that the hybrid algorithms we instantiate are able to match and improve over the best classical SLS method.

1 Introduction

Simple and hybrid stochastic local search (SLS) algorithms [13] have received an enormous attention over the last two decades. Often, one of the crucial research questions is which of the many different general-purpose SLS methods (also known as metaheuristics) is the most suited for a specific type of problem and how much performance improvement can be further obtained by considering the hybridization of such methods. In the early stages of the research on SLS methods, these questions were tackled across various papers. Roughly speaking, in each paper an SLS methods was adapted to the specific problem under concern and the computational results were compared to those of other papers. The potential pitfalls of this "horse-race" kind of research have been described in various occasions [12,16]. A better approach may be to run careful experimental

M.J. Blesa, C. Blum, and S. Voß (Eds.): HM 2014, LNCS 8457, pp. 30–44, 2014.
© Springer International Publishing Switzerland 2014

comparisons of algorithms in a same controlled environment as done in more recent works [22,25,27]. Such an approach avoids some pitfalls, such as comparing algorithms across different computers and implementation languages, uneven implementation effort and expertise, and data structures, which leads to a high noise in the comparisons. This approach is certainly very useful in settling the state of knowledge for what concerns the solution of specific problems by SLS. However, the SLS algorithms compared are typically re-implementations of the originally proposed algorithms and they use the originally proposed parameter settings; therefore, these papers also inherit potential pitfalls in the original papers such as uneven parameter tuning or sub-optimal algorithm designs.

In this paper, we follow a different approach for comparing SLS methods and evaluating the impact hybrid SLS methods have over single ones. We propose to use automatically configurable SLS algorithm frameworks for performing algorithm comparisons and studying the impact of hybridization. This approach has a number of advantages. In particular, various low level heuristics can be used by all the automatically configured SLS methods. In fact, it is known that for the engineering of effective SLS algorithms, the underlying low-level heuristics are decisive for reaching high performance. Giving the possibility of using the same algorithmic components should also contribute towards reducing the variance of computational results. Automatic configuration of the SLS algorithms also helps to make the performance less dependent on the algorithm designers intuition and ensures a comparable tuning effort.

In this article, we use a configurable algorithmic framework for the automatic generation of SLS algorithms. This framework was introduced in [21] and it allows the automatic generation of simple and hybrid SLS algorithms that essentially manipulate a single solution at each step. In particular, the framework is based on a recursive template of an iterated local search (ILS) [20] that allows the instantiation of (i) SLS methods such as randomized iterative improvement (RII) [28,13], probabilistic iterative improvement (PII) [13], simulated annealing (SA) [18,2], tabu search (TS) [10], ILS [20], greedy randomized adaptive search procedures (GRASP) [7], iterated greedy (IG) [29], and variable neighbourhood search (VNS) [11]; and (ii), many different ways of hybridizing these SLS methods, through having different methods at each level of the recursive template. In this paper, we study the generation and comparison of simple SLS algorithms and hybrids thereof, using as example problems two permutation flow-shop scheduling problems, the total weighted tardiness (PFSP-WT) and the total weighted completion times flow-shop problems (PFSP-WCT).

The paper is structured as follows. In Section 2 we give an overview of the framework for the automatic generation of SLS algorithms; in Section 3 we describe the two optimisation problems used for comparing the generated algorithms; in Section 4 we present the experimental evaluation and the results; and finally in Section 5 we summarize the conclusions.

2   Framework

Methods for automatic design of algorithms are often based on instantiating algorithms from a grammar using evolutionary algorithms [8,1] or automatically configuring the parameters of flexible algorithmic frameworks [17,19]. We have recently proposed a method for the automatic generation of heuristics based on grammars but using automatic algorithm configuration tools [23]. Our method starts from a grammar defining the algorithmic building blocks and how they can be combined. From this grammar, we first generate a parameter space that maps the derivation rules defined in the grammar, and then we use an automatic algorithm configuration tool to search the algorithmic design space for an effective algorithm. One of the main advantages of our method is that it combines the flexibility of grammars for defining the algorithm design space with the effectiveness of tools for automatic algorithm configuration. The method was successfully applied to the generation of iterated greedy algorithms for the permutation flowshop problem with weighted tardiness objective [23], and one dimensional bin-packing [24].

Later, we extended our method in order to generate more types of simple and hybrid SLS algorithms [21]. To do so, we defined, in terms of a problem-independent grammar, the algorithmic building blocks of a recursive iterated local search template using the ParadisEO [14] local search framework. From such a *generalised local search* template, one can instantiate several SLS methods known in the literature as well as hybridizations thereof. The flexibility of the grammar allows, for example, the instantiation of a simulated annealing or an iterated greedy algorithm as well as more complex hybrids such as an iterated local search that has as subsidiary local search a VNS, which in turn uses simulated annealing as the local search. Moreover, in a second, separate grammar, we define the algorithmic building blocks to generate problem-specific heuristics. The procedure is completely automated: once the problem-specific grammar is defined, an automated system integrates this second grammar with the problem-independent one, it generates the parameters to describe the alternative choices in the algorithmic design space defined in the grammar, and, given a set of training instances, it searches for an effective algorithm for the problem at hand, by automatically generating, compiling, testing, and evaluating the performance of tenths of thousands of candidate SLS algorithms until a computational budget is exhausted [21].

Figures 1, 2, and 3 show a subset of the derivation rules of a simplified version of the problem-independent grammar. The rules in Fig. 1 define the target algorithm <algo_choice> as the aforementioned generalised local search procedure <GLS> or one of a series of known SLS methods <known_MH>. The generalised local search procedure is defined as a recursive iterated local search, which is specified by a perturbation, an acceptance criterion, and, as subsidiary local search, an <algo_choice>. If the latter is again a <GLS>, the rule is recursively evaluated and a new iterated local search algorithm is defined.

The <GLS> procedure could per se be instantiated into any of the known SLS methods defined in <known_MH>. In fact, as shown in Table 1, such known SLS

```
          <start> ::= <problem_specific_initial_solution> <algo_choice>
    <algo_choice> ::= <known_MH> | <GLS>
       <known_MH> ::= <SA> | <ILS> | <PII> | <RII> | <IG> | <VNS> | <TS>
            <GLS> ::= <perturbation> <acceptance> <algo_choice>
```

Fig. 1. Grammar description of the generalised local search procedure

```
           <SA> ::= <perturbation_one_move> <ls_none> <accept_metropolis>
          <ILS> ::= <perturbation> <local_search> <accept_ILS>
          <PII> ::= <perturbation_one_move> <ls_none> <accept_metr_fix_temp>
          <RII> ::= <perturbation_one_move> <ls_none> <accept_probab>
          <VNS> ::= <perturb_VNS> <local_search> <accept_ILS>
           <IG> ::= <problem_specific_destruct_reconstruct>
                    <local_search> <acceptance>
           <TS> ::= <perturbation_none> <local_search_TS> <accept_always>
   <accept_ILS> ::= <accept_always> | <accept_better> | <accept_better_equal>
 <perturb_VNS> ::= <pert_strength_dyn_incr> <repeatable_perturb>
```

Fig. 2. Known SLS methods that can be instantiated from the grammar

```
    <pert_strength> ::= <fixed_str> | <random_str> | <dynamic_inc> | <dynamic_dec>
     <perturbation> ::= <pert_restart> | <kopt_pert> | <kopt_pert_stop>
                      | <pert_none> | <repeatable_pert> | <problem_specific_pert>
   <local_searches> ::= <ls> (<comparator>)
              <ls> ::= <first_imp_hc_ls> | <best_imp_hc_ls> | <ls_none>
                      | <first_imp_cont_hc> | <problem_specific_ls>
       <comparator> ::= <better> | <better_equal>
  <temp_relative_to> ::= <best> | <current>
 <cooling_schedule> ::= <rel_fix_temp> | <rel_schedule> | <rel_schedule_stop_at_min>
                      | <rel_schedule_reheat>
       <acceptance> ::= <accept_always> | <accept_better> | <accept_better_equal>
                      | <accept_probab> | <accept_threshold> | <accept_random>
                      | <accept_metropolis> | <accept_metr_fix_temp>
```

Fig. 3. Other significant derivation rules in the grammar. Local searches, and pertubations can be extended by defining them in the problem-specific grammar.

methods can be seen as particular cases of our generalised local search template. The reason why these SLS methods are also defined with their own derivation rules is that in this way we increase the probability that such specific combinations of algorithmic building blocks are evaluated by the tool for automatic algorithm configuration used to search the algorithmic design space.

Figure 2 shows a subset of the simplified rules for the known SLS methods defined in <known_MH>. For example, the probabilistic iterative improvement <PII> is defined as an iterative improvement method in which the perturbation makes a single move in the neighbourhood <perturbation_one_move>, there is no local search <ls_none>, and the acceptance criterion is a Metropolis condition

Table 1. Classical SLS algorithms modeled after the GLS scheme

Name	Perturbation	Local Search	Acceptance Criterion
SA [18]	one move	∅	Metropolis
PII [13]	one move	∅	Metropolis (fixed temp.)
RII [28,13]	one move	∅	probability
VNS [11]	variable move	first-improv. descent	improvingStrictly
IG [29]	deconstruction-construction	"any"	"any"
TS [10]	∅	tabu search	always accept

with a fixed temperature. Some of the derivation rules such as `<accept_perturb>` limit the range of possible values assumed by a building block, in this case, the derivation limits the acceptance criterion to three criteria commonly used in ILS algorithms. Figure 3 shows other sample derivations in the grammar for the local searches, perturbations, and acceptance criteria. All derivations starting with `<problem_specific_...>` are derivations defined in the problem-specific grammar and are automatically integrated in the problem-independent one.

By using the grammar described above or subsets of it, we can generate either hybrid SLS algorithms, or just limit ourselves to generate well-known non-hybrid SLS algorithms, or even just focus on a single SLS method and tune its numerical parameters, which are not shown in the grammars for brevity. In this paper, we use various subsets of the grammars described here to investigate the generation and tuning of hybrid versus non-hybrid SLS algorithms.

3 The Permutation Flowshop Problem

As a benchmark problem, we consider the widely studied permutation flowshop scheduling problem (PFSP) [15]. The PFSP is an $\mathcal{NP}$-hard [9] problem, thus tackling large instances often requires the use of heuristic algorithms. Moreover, the PFSP models a very common setup in industrial production, thus, it is of practical relevance. For these reasons, the PFSP is still an important benchmark problem for the design and comparison of SLS methods. There are many formulations of the PFSP focusing on different objectives and with various constraints. When tackling less-studied variants, the use of automatic generation to design new algorithms can save a significant amount of effort.

The goal in the flowshop problem (FSP) is to schedule a set of n jobs $(J_1, \ldots, J_n)$ on m machines $(M_1, \ldots, M_m)$, and all jobs must be processed on the machines in the same order, i.e., all jobs have to be processed on machine M_1, then machine M_2, and so on until machine M_m. A common restriction in the FSP is to forbid job passing between machines, i.e., to restrict to solutions that are permutations of jobs, resulting in the PFSP. In the PFSP, all processing times p_{ij} for a job J_i on a machine M_j are fixed, known in advance, and non-negative. Formally, the PFSP consists of finding a job permutation π, where π_i denotes the job in the i-th position, that minimizes a given objective function $F(\pi)$

$$\min \quad F(\pi)$$
$$\text{subject to} \quad C_{\pi_0 j} = 0 \quad j \in \{1, \ldots, m\},$$
$$C_{\pi_i 0} = 0 \quad i \in \{1, \ldots, n\}, \tag{1}$$
$$C_{\pi_i j} = \max\{C_{\pi_{i-1} j}, C_{\pi_i j-1}\} + p_{ij},$$
$$i \in \{1, \ldots, n\} \quad j \in \{1, \ldots, m\},$$

where C_{ij} denotes the completion time of a job i on machine j. Depending on the definition of $F(\pi)$, there are many variants of the FPSP. We study here the following two variants.

Minimization of the Sum of Weighted Completion Times (PFSP-WCT). In many practical situations, some jobs are more important than others, which is represented by a weight associated to them. In such a case, the objective may be to minimize the time required to complete each job on the last machine, weighted by their importance:

$$\min \quad F(\pi) = \sum_{i=1}^{n} w_i \cdot C_i \tag{2}$$

where C_i denotes the completion time of a job i on the last machine, and w_i is the weight assigned to job J_i to specify its relative importance. The PFSP-WCT is strongly $\mathcal{NP}$-hard already for two machines [9].

Minimization of the Total Weighted Tardiness (PFSP-WT). In other practical scenarios, e.g., when products are due to customers at a specific time, jobs have an associated *due date* (d_i for a job J_i), and the *tardiness* of a job J_i is defined as $T_i = \max\{C_i - d_i, 0\}$. In such a case, the goal may be to minimize the total tardiness, weighted by the relative importance of jobs:

$$\min \quad F(\pi) = \sum_{i=1}^{n} w_i \cdot T_i. \tag{3}$$

The PFSP-WT is $\mathcal{NP}$-hard in the strong sense even for a single machine [4].

3.1   Problem-Specific Components for the PFSP

The problem-specific components of the grammar (Fig. 3) in the case of the PFSP define constructive heuristics for generating the initial solutions, iterative improvement algorithms, a perturbative search, the computation of the objective function, and other helper functions to represent the problem instance in memory in a format suitable for being used within ParadisEO.

For both, the PFSP-WT and the PFSP-WCT problem, the constructive heuristics implemented in our framework are a random permutation and an NEH [26] heuristic, which is an insertion heuristic known from the PFSP with makespan objective. For the PFSP-WT problem, there is also a SLACK heuristic for computing an initial ordering of the jobs to be passed to the NEH heuristic [5]. Further components define single moves in the swap, exchange, and insert neighborhoods, and a random perturbative search operator, in which k jobs are removed randomly from the current solution and reinserted in an order which minimises the objective value of the partial solution. The iterative improvement local search procedure, is a first-improvement algorithm in a swap neighborhood, with a maximum number of swaps corresponding to $2 \cdot (n - 1)$, where n is the number of jobs.

4 Experimental Evaluation

In this paper, we address two main questions related to the automatic generation of hybrid SLS algorithms. The first question asks whether it is more advantageous to tune all classical (non-hybrid) SLS methods together or to tune them separately. Once we found the best way to tune the classical SLS methods, the second question concerns whether there is an advantage on spending the total computing budget searching for a hybrid SLS method, or it is more effective to spend the same budget on selecting and tuning classical (non-hybrid) SLS methods.

4.1 Experimental Setup

We consider the following grammar descriptions:

- Individual classical SLS methods (PII, RII, SA, ILS, IG, VNS, TS). In this case, the design is fixed, and the task is to find the best parameter settings of each individual SLS method. We can generate the grammar for automatically configuring specific SLS methods by (automatically) adapting the rules given in Fig. 1. In particular, it suffices to change the first rule `<start> ::= <problem_specific_initial_solution> <algo_choice>` to a new rule `<start> ::= <problem_specific_initial_solution> <SLS-M>` , where `<SLS-M>` is any of the classical SLS methods we would like to include such as PII, RII, SA, etc. As a result, the rule for `<algo_choice>`, `<known_MH>`, `<GLS>`, as well as all other rules that cannot be reached by a sequence of derivation from `<start>`, are implicitly removed, and the number of parameters (reported in Table 2) depends on the specific SLS method tuned.
- MH. This grammar corresponds to the combination of all individual SLS methods. In this case, the task is to select the best individual SLS method and tune its parameter settings. This can be done by changing the rule `<algo_choice> ::= <known_MH> | <GLS>` in Fig. 1 to `<algo_choice> ::= <known_MH>`. This grammar generates 107 parameters in total, which includes the parameters to model the choice among the various SLS methods and all the parameters that arise in each of the specific SLS methods.

Table 2. Total number of parameters for each strategy

Hybrid	MH	PII	RII	SA	ILS	IG	VNS	TS
286	107	4	2	27	27	40	10	2

- Hybrid. This grammar corresponds to adding the possibility of hybridization to MH, that is, the grammar can instantiate individual SLS methods as well as hybridizations of them. This can be done by changing the rule `<algo_choice> ::= <known_MH> | <GLS>` to `<algo_choice> ::= <GLS>`. We limit the depth of the recursive rule `<GLS>` to three, thus at most three hybridization levels are allowed. This grammar generates 286 parameters in total.

Our experiments consist on applying irace, a tool for automatic algorithm configuration, to the different grammar descriptions above in order to find the best instantiation of the grammar, i.e., the best algorithm configuration and parameter setting, given a set of training problem instances and a tuning budget [23]. The tuning budget is defined as a maximum number of runs of the generated algorithms. Hence, each algorithm that is genereated from the grammar corresponds to one configuration tested during a run of irace. The best configuration found by irace is then applied to a set of test instances, different from the training set, in order to assess its performance.

For each problem, the PFSP-WT and the PFSP-WCT, we generate 15 random instances for each of $n = \{50, 60, 70, 80, 90, 100\}$ jobs and $m = 20$ machines. Five instances are used for training, and the other ten for testing. The algorithms generated by irace are run 30 times, with different random seeds, on the ten test instances. Since the tuning process is stochastic, we repeat the aforementioned tuning/testing procedure three times to account for the variability of the tuning process.

The tuning budget considered here for each run of irace is 50 000 runs of an algorithm, and each run of any of the specific algorithm configuration generated from the grammar is stopped after 20 CPU-seconds. Thus, one run of irace requires roughly ten hours when running on a cluster of 30 AMD Opteron 6272 2.1 GHz CPU-cores running on CentOS 6.2 Linux, that is, a single automatic configuration run can be done overnight (while sleeping). The process of instantiating an algorithm from a grammar produces C++ code that is compiled with GCC 4.7.2 with options "`-Ofast -flto -march=native`". In the case of individual SLS methods, we divide the total tuning budget equally among the seven SLS methods (PII, RII, SA, ILS, IG, VNS, TS), i.e., 7 143 runs for each SLS method.

4.2   Experimental Results for PFSP-WCT

As described above, we first apply irace three times to the various grammars. For the PFSP-WCT, the best configurations found by irace are rather consistent when considering the MH grammar, where the selected SLS method is always VNS. In the case of Hybrid, it is either (i) an ILS with Metropolis condition as acceptance criterion, a cooling schedule, random swap moves as perturbation, and first-improvement as local search, or (ii) a more complex hybridisation with two or three levels of hybridisation.

We run each of the algorithms generated by irace on the test instances and rank them per instance (the lower the rank, the better). We also apply a Friedman test [3] to determine what difference of ranks is statistically significant. The results are shown in Table 3. They indicate that the hybrid algorithm is consistently better than the other tuned algorithms. The second ranked algorithms is MH. As mentioned above, MH corresponds to a VNS, which is better tuned than the stand-alone VNS, because of the difference in available budget. In fact, when tuned as MH, irace has the possibility of allocating more tuning effort to the best performing SLS methods. When tuning a standalone VNS, the allocated budget is 1/7-th of the total tuning budget, instead.

Table 3. Comparison of the strategies through the Friedman test blocking on the instances of the PFSP-WCT. $\Delta R_{\alpha=0.05}$ gives the minimum difference in the sum of ranks between two strategies that is statistically significant.

$\Delta R_{\alpha=0.05}$	Strategy (ΔR)
28.53	**Hybrid** (0), MH (40), SA (46), VNS (75), ILS (138), IG (139), RII (276), PII (342), TS (402)

In terms of solution quality, the differences are not very large, as shown by the mean relative percentage deviations from the best-found solutions given in Table 4. The only notable exception is TS, which performs substantially worse than the rest. This might be due to our implementation of the tabu list in ParadisEO [14] that stores the complete solutions in the tabu list. More efficient schemas storing the moves in the neighbourhood would allow for a larger number of iterations and better results in the 20 CPU-seconds allowed to the algorithm.

Table 4. Relative percentage deviations from the best known solutions averaged over three tuning experiments, 30 runs, and ten instances of the PFSP-WCT

Instance size	Hybrid	MH	PII	RII	SA	ILS	IG	VNS	TS
50x20	0.25	0.25	0.70	0.42	0.32	0.27	0.26	0.26	2.95
60x20	0.54	0.55	0.93	0.75	0.58	0.57	0.56	0.57	3.38
70x20	0.77	0.75	1.26	1.00	0.74	0.82	0.84	0.78	3.66
80x20	0.78	0.81	1.27	1.01	0.73	0.84	0.85	0.81	3.68
90x20	0.84	0.85	1.42	1.06	0.73	0.94	0.95	0.87	3.76
100x20	0.86	0.90	1.60	1.05	0.74	0.98	0.98	0.93	3.74

Given the fact that we use a heterogeneous set of instances with various sizes, we could expect some variation on the best algorithm for each instance size. In Table 5, we partition the results per instance size. Although there is some variability on the rankings, due to the effect of the instance size, overall, Hybrid, MH, SA, and VNS are ranked consistently high across all instance sizes. To further analyse the overall robustness of the automatic generation method, we present in Table 6 the ranks of the algorithm across the three separate tuning experiments. Also in this case, the best algorithms in Table 3 are also the highest ranked ones in the single tuning experiments.

Table 5. Comparison of the strategies on the single instance sizes for the PFSP-WCT. The ranks are given in parentheses.

Instance size	Strategy (ΔR)
50x20	MH (0), Hybrid (3), IG (5), ILS (13), VNS (15), SA (36), RII (47), PII (57), TS (67)
60x20	Hybrid (0), MH (10), IG (15), VNS (16), ILS (20), SA (29), RII (50), PII (60), TS (70)
70x20	Hybrid (0), SA (1), MH (3), VNS (7), ILS (22), IG (33), RII (46), PII (56), TS (66)
80x20	Hybrid (0), SA (4), VNS (15), MH (15), ILS (24), IG (26), RII (49), PII (59), TS (69)
90x20	SA (0), Hybrid (9), MH (14), VNS (17), IG (36), ILS (39), RII (53), PII (64), TS (74)
100x20	SA (0), Hybrid (12), MH (22), VNS (29), ILS (44), IG (48), RII (55), PII (70), TS (80)

Table 6. Comparison of the strategies obtained with three different tuning for the PFSP-WCT. The ranks are computed across all instance sizes.

Tuning	Strategy (ΔR)
irace 1	SA (0), Hybrid (15), MH (27), ILS (66), VNS (76), IG (119), RII (252), PII (317), TS (379)
irace 2	MH (0), VNS (30), Hybrid (39), SA (53), IG (101), ILS (130), RII (256), PII (327), TS (387)
irace 3	Hybrid (0), SA (31), VNS (87), MH (89), IG (122), ILS (155), RII (273), PII (347), TS (408)

4.3   Experimental Results for PFSP-WT

In the case of the PFSP-WT, similarly to the results for PFSP-WCT, irace selected two VNS algorithms and one IG algorithm when using the grammar described by MH. In the case of Hybrid, irace generated one hybrid algorithm with two levels of hybridization and two ILS algorithms with a Metropolis condition as acceptance criterion.

Table 7 shows the rankings among the tuned algorithms and the statistically significant differences according to the Friedman test. Also in this case, the hybrid algorithm performs significantly better than the rest; the second highest ranked algorithm is MH followed by IG, and ILS. MH has the advantage of tuning all algorithms at the same time, hence to dynamically invest more tuning budget on the most promising algorithmic schemas from the beginning of the tuning. IG and ILS are also highly ranked. This comes at no surprise; ILS is the algorithm that was often selected when tuning Hybrid, and the state of the art for the PFSP-WT is an IG algorithm [29,6].

Table 7. Comparison of the strategies through the Friedman test blocking on the instances of the PFSP-WT. $\Delta R_{\alpha=0.05}$ gives the minimum difference in the sum of ranks between two strategies that is statistically significant.

$\Delta R_{\alpha=0.05}$	Strategy (ΔR)
28.75	**Hybrid** (0), MH (50), IG (54), ILS (54), VNS (59), RII (222), SA (236), PII (333), TS (396)

Also for the PFSP-WT problem, in most cases, the difference between the tuned algorithms in terms of average percentage deviation from the best known solutions are minor (Table 8). For what concerns the robustness of the automatic generation method, Hybrid, MH, IG, and ILS are consistently among the highest ranked algorithms on the single instance sizes (Table 9) as well as across the three different tuning experiments (Table 10).

Overall the results suggest that the use of a flexible grammar like the one of Hybrid, allows to efficiently invest the tuning budget in the promising schemas and obtain an effective simple or hybrid SLS algorithm for the problem at hand. For both problem studied, complex hybridisations of the SLS methods were shown to be among the top performing algorithms. However, when considering relative percentage deviation, the difference to the best performing tuned known SLS methods are small, although statistically significant.

Table 8. Relative percentage deviations from the best known solutions averaged over three tuning experiments, 30 runs, and ten instances of the PFSP-WT

Instance size	Hybrid	MH	PII	RII	SA	ILS	IG	VNS	TS
50x20	2.67	2.75	17.33	4.68	7.87	3.98	2.12	3.01	26.06
60x20	2.02	2.11	6.06	3.37	3.66	2.13	1.97	2.20	15.16
70x20	2.79	2.93	5.44	4.15	4.33	3.05	2.90	2.94	14.47
80x20	2.95	2.93	4.70	3.75	3.35	2.97	3.04	3.04	13.33
90x20	2.71	2.80	4.26	3.37	3.23	2.68	2.91	2.73	11.26
100x20	2.87	2.91	5.49	3.31	4.35	2.84	3.20	2.84	10.84

Table 9. Comparison of the strategies on the single instance sizes for the PFSP-WT. The ranks are given in parentheses.

Instance size	Strategy (ΔR)
50x20	IG (0), Hybrid (12), MH (13), VNS (26), ILS (32), RII (43), SA (57), PII (65), TS (76)
60x20	IG (0), Hybrid (5), MH (13), ILS (14), VNS (18), RII (42), SA (48), PII (60), TS (70)
70x20	Hybrid (0), VNS (9), IG (14), MH (15), ILS (22), RII (44), SA (51), PII (61), TS (72)
80x20	MH (0), Hybrid (2), ILS (6), IG (13), VNS (15), SA (24), RII (45), PII (55), TS (65)
90x20	ILS (0), Hybrid (3), VNS (11), IG (16), MH (17), SA (30), RII (43), PII (56), TS (67)
100x20	Hybrid (0), ILS (2), VNS (2), MH (14), RII (27), IG (33), SA (48), PII (58), TS (68)

Table 10. Comparison of the strategies obtained with three different tuning for the PFSP-WT. The ranks are computed across all instance sizes.

Tuning	Strategy (ΔR)
irace 1	MH (0), Hybrid (23.5), ILS (33), VNS (46.5), IG (73), RII (188), SA (238), PII (310), TS (384)
irace 2	IG (0), ILS (4.5), Hybrid (33.5), VNS (40), MH (78.5), RII (198.5), SA (275.5), PII (306.5), TS (372.5)
irace 3	Hybrid (0), MH (35), IG (38), VNS (54), ILS (96), SA (124), RII (227), PII (316), TS (370)

5 Discussion and Conclusions

We have examined the design of hybrid SLS algorithms in comparison with the selection and tuning of non-hybrid SLS methods. In order to perform a unbiased comparison, the design, selection and tuning is carried out automatically using an automatic design framework that was recently proposed. This framework is composed of a grammar description, from which various SLS methods and hybrids thereof may be instantiated, and the application of an automatic algorithm configuration method (irace), which searches for the best instantiation of the grammar given a set of training instances and a tuning budget.

In this paper, we generated algorithms in three different ways: (*i*) equally distributing the budget into the independent tuning of the parameters of classical SLS methods; (*ii*) spending all the budget on selecting the best classical SLS method at the same time as the parameter tuning is performed; and (*iii*) generating directly hybrid SLS algorithms that may reproduce the classical SLS methods and hybridizations thereof.

We carried out experiments on two variants of the PFSP, concretely, the minimization of the weighted sum of completion times (PFSP-WCT) and the minimization of the weighted tardiness (PFSP-WT). For both problems, method (*ii*) is always able to identify one of the best SLS methods, which are VNS, IG and ILS. The observed differences between these three SLS methods seem to be more related to sufficiently tuning their parameters than to the choice of SLS method. This is not entirely unexpected, since IG is the state-of-the-art algorithm for these two problems [29,6], whereas VNS and ILS combine perturbation moves with local search.

Overall, rather than tuning the non-hybrid SLS methods separately, it seems more advantageous to let the automatic method to choose the most appropriate SLS method and tune it at the same time, that is, to follow the grammar MH described in the paper. Moreover, by allowing recursive rules in the grammar, the automatic method is able to generate more complex hybrid SLS algorithms when the problem requires it.

Method (*iii*) generates a variant of ILS with a Metropolis acceptance criterion or hybrid ILS algorithms with at least two levels of hybridization. For both problems, the difference between the hybrid algorithm and the other algorithms tuned are always statistically significant. This result shows the flexibility of method (*iii*), which is able to adapt the complexity of the generated hybrid SLS algorithm to the problem and generate effective algorithms.

The methodology presented here can be straightforwardly extended to other problems for which the use of hybrid SLS methods is appropriate. Moreover, we are currently working on extending the proposed framework to include additional SLS methods and allow other forms of hybridization. Finally, this paper only analyzes the performance of the algorithms generated at the end of the automatic design process, whereas it would be interesting to collect additional information during the process. We are currently investigating which information should be collected in order to obtain other useful conclusions from the automatic design process.

Acknowledgments. This work was supported by the COMEX project within the Interuniversity Attraction Poles Programme of the Belgian Science Policy Office. Franco Mascia, Manuel López-Ibáñez, and Thomas Stützle acknowledge support from the Belgian F.R.S.-FNRS, of which they are post-doctoral researchers and a senior research associate, respectively.

References

1. Burke, E.K., Hyde, M.R., Kendall, G.: Grammatical evolution of local search heuristics. IEEE Transactions on Evolutionary Computation 16(7), 406–417 (2012)
2. Cerný, V.: A thermodynamical approach to the traveling salesman problem. Journal of Optimization Theory and Applications 45(1), 41–51 (1985)
3. Conover, W.J.: Practical Nonparametric Statistics, 3rd edn. John Wiley & Sons, New York (1999)

4. Du, J., Leung, J.Y.T.: Minimizing total tardiness on one machine is NP-hard. Mathematics of Operations Research 15(3), 483–495 (1990)
5. Dubois-Lacoste, J.: A study of Pareto and Two-Phase Local Search Algorithms for Biobjective Permutation Flowshop Scheduling. Master's thesis, IRIDIA, Université Libre de Bruxelles, Belgium (2009)
6. Dubois-Lacoste, J., López-Ibáñez, M., Stützle, T.: A hybrid TP+PLS algorithm for bi-objective flow-shop scheduling problems. Computers & Operations Research 38(8), 1219–1236 (2011)
7. Feo, T.A., Resende, M.G.C.: Greedy randomized adaptive search procedures. Journal of Global Optimization 6, 109–113 (1995)
8. Fukunaga, A.S.: Automated discovery of local search heuristics for satisfiability testing. Evolutionary Computation 16(1), 31–61 (2008)
9. Garey, M.R., Johnson, D.S., Sethi, R.: The complexity of flowshop and jobshop scheduling. Mathematics of Operations Research 1, 117–129 (1976)
10. Glover, F.: Tabu search – Part I. INFORMS Journal on Computing 1(3), 190–206 (1989)
11. Hansen, P., Mladenovic, N.: Variable neighborhood search: Principles and applications. European Journal of Operational Research 130(3), 449–467 (2001)
12. Hooker, J.N.: Testing heuristics: We have it all wrong. Journal of Heuristics 1(1), 33–42 (1996)
13. Hoos, H.H., Stützle, T.: Stochastic Local Search—Foundations and Applications. Morgan Kaufmann Publishers, San Francisco (2005)
14. Humeau, J., Liefooghe, A., Talbi, E.G., Verel, S.: ParadisEO-MO: From fitness landscape analysis to efficient local search algorithms. Journal of Heuristics 19(6), 881–915 (2013)
15. Johnson, D.S.: Optimal two- and three-stage production scheduling with setup times included. Naval Research Logistics Quarterly 1, 61–68 (1954)
16. Johnson, D.S.: A theoretician's guide to the experimental analysis of algorithms. In: Goldwasser, M.H., et al. (eds.) Data Structures, Near Neighbor Searches, and Methodology: Fifth and Sixth DIMACS Implementation Challenges, pp. 215–250. American Mathematical Society, Providence (2002)
17. KhudaBukhsh, A.R., Xu, L., Hoos, H.H., Leyton-Brown, K.: SATenstein: Automatically building local search SAT solvers from components. In: Boutilier, C. (ed.) Proceedings of the Twenty-First International Joint Conference on Artificial Intelligence (IJCAI 2009), pp. 517–524. AAAI Press, Menlo Park (2009)
18. Kirkpatrick, S., Gelatt, C.D., Vecchi, M.P.: Optimization by simulated annealing. Science 220, 671–680 (1983)
19. López-Ibáñez, M., Stützle, T.: The automatic design of multi-objective ant colony optimization algorithms. IEEE Transactions on Evolutionary Computation 16(6), 861–875 (2012)
20. Lourenço, H.R., Martin, O., Stützle, T.: Iterated local search: Framework and applications. In: Gendreau, M., et al. (eds.) Handbook of Metaheuristics, ch. 9, 2nd edn. International Series in Operations Research & Management Science, vol. 146, pp. 363–397. Springer, New York (2010)
21. Marmion, M.E., Mascia, F., López-Ibáñez, M., Stützle, T.: Automatic design of hybrid stochastic local search algorithms. In: Blesa, M.J., Blum, C., Festa, P., Roli, A., Sampels, M. (eds.) HM 2013. LNCS, vol. 7919, pp. 144–158. Springer, Heidelberg (2013)
22. Martí, R., Reinelt, G., Duarte, A.: A benchmark library and a comparison of heuristic methods for the linear ordering problem. Computational Optimization and Applications 51(3), 1297–1317 (2012)

23. Mascia, F., López-Ibáñez, M., Dubois-Lacoste, J., Stützle, T.: From grammars to parameters: Automatic iterated greedy design for the permutation flow-shop problem with weighted tardiness. In: Nicosia, G., Pardalos, P. (eds.) LION 7. LNCS, vol. 7997, pp. 321–334. Springer, Heidelberg (2013)
24. Mascia, F., López-Ibáñez, M., Dubois-Lacoste, J., Stützle, T.: Grammar-based generation of stochastic local search heuristics through automatic algorithm configuration tools. Tech. Rep. TR/IRIDIA/2013-015, IRIDIA, Université Libre de Bruxelles, Belgium (2013)
25. Minella, G., Ruiz, R., Ciavotta, M.: A review and evaluation of multiobjective algorithms for the flowshop scheduling problem. INFORMS Journal on Computing 20(3), 451–471 (2008)
26. Nawaz, M., Enscore Jr., E., Ham, I.: A heuristic algorithm for the m-machine, n-job flow-shop sequencing problem. OMEGA 11(1), 91–95 (1983)
27. Pan, Q.K., Ruiz, R.: Local search methods for the flowshop scheduling problem with flowtime minimization. European Journal of Operational Research 222(1), 31–43 (2013)
28. Papadimitriou, C.H., Steiglitz, K.: Combinatorial Optimization – Algorithms and Complexity. Prentice Hall, Englewood Cliffs (1982)
29. Ruiz, R., Stützle, T.: A simple and effective iterated greedy algorithm for the permutation flowshop scheduling problem. European Journal of Operational Research 177(3), 2033–2049 (2007)

A Local Search Approach for Binary Programming: Feasibility Search

Samuel Souza Brito, Haroldo Gambini Santos,
and Bruno Henrique Miranda Santos

Computing Department,
Universidade Federal de Ouro Preto (UFOP)
Ouro Preto – Minas Gerais - Brasil
samuelsouza,haroldo@iceb.ufop.br,
bruno_h_m_s@hotmail.com

Abstract In this paper we propose a Local Search approach for NP-Hard problems expressed as binary programs. Our search method focuses on the fast production of feasible solutions. The method explicitly considers the structure of the problem as a conflict graph and uses a systematic neighbor generation procedure to jump from one feasible solution to another using chains of movements. Computational experiments comparing with two open source integer programming solvers, CBC and GLPK, in MIPLIB 2010 instances, showed that our approach is more reliable for the production of feasible solutions in restricted amounts of time.

Keywords: Binary Programming, Heuristics, Local Search.

1   Introduction

In this work we consider Binary Linear Programs, or Binary Programs (BP), which can be expressed as:

$$min. :$$
$$c^T x \tag{1}$$
$$s.t. :$$
$$l \le Ax \le u \tag{2}$$
$$x_j \in \{0, 1\} \ \forall j \in J \tag{3}$$

Where x is a vector of n binary variables with its associated cost vector c to be minimized (1). A is a matrix with dimension $m \times n$ expressing the constraint system where each constraint has a lower and upper bound expressed in vectors l and u respectively.

In spite of its simplicity, Binary Programming , is one of the most important techniques in Operations Research (OR). Some notable applications include The Traveling Salesman Problem [1], Project Scheduling [2] and Computational Biology [3]. The availability of constantly improving optimization packages [4], some

M.J. Blesa, C. Blum, and S. Voß (Eds.): HM 2014, LNCS 8457, pp. 45–55, 2014.
© Springer International Publishing Switzerland 2014

of which are open source [5], has made Binary Programming a great choice for OR practitioners.

Linear Programming based methods for Binary Programming typically work by solving series of linear programs using some variation of the Branch-and-Bound method [6]. This method works with fractional solutions but the systematic exploration of a tree of progressively restricted subproblems eventually produces an Integer Feasible solution. Since the fractional solution is usually useless for practical purposes, solvers are also being evaluated [7] considering their ability to quickly produce an Integer Feasible solution.

In order to obtain feasible solutions in acceptable computational time, this paper presents a hybrid heuristic. This approach is characterized by two phases: a constructive phase, which involves solving the maximum independent set problem and a local search phase. Both phases work with information provided by a conflict graph, created from the analysis of the constraints imposed by the problem input. They do not require a black-box linear solver or branch-and-bound family methods. Experiments with binary problems of the Mixed Integer Programming Library (MIPLIB) 2010 show that the proposed approach is able to produce more feasible solutions than the GNU Linear Program Kit (GLPK) and COIN-OR [8] CBC [9] solvers in restricted time intervals.

The paper is organized as follows: Section 2 presents related works, Section 3 presents a description of the proposed approach. In Section 4 computational experiments with MIPLIB are presented. Finally, Section 5 discusses future works and conclusions.

2   Related Works

One common approach to the development of heuristics for Integer Programming is to use the information from the linear programming relaxation alone or combined with some black-box integer programming solver. These methods solve series of linear programs iteratively [10–12]. One of them, called Feasibility Pump, is a smart and simple heuristic, proposed by Fischetti et al. [10]. The purpose of this heuristic is to find an initial feasible solution, even in difficult problems. The basic idea is to start with a relaxed linear solution and then make changes in the objective function to try to minimize the infeasibility related to the integrality constraints. So, this method is designed to pump the feasibility of a relaxed solution for an integer solution. Performed tests show that this method is able to find feasible solutions quickly and can be used in other methods to accelerate the search process.

Based in structure of Pure Integer Programming problems, in [13] is proposed a approach that uses a genetic algorithm. In this approach, a gene corresponds to a decision variable of the problem, represented by a bit array. Therefore, a chromosome is defined by a decision variable set, which the fitness is the objective function. The initial population is generated randomly, respecting the variables domains defined by constraints. Then, the method attempts to remove the infeasibilities exchanging the variable values by rounded values of the linear relaxation of the problem. The basic steps of a genetic algorithm are performed: crossover, mutation and selection. Crossover uses an one point operator. Mutation is made inverting a bit value selected randomly of a chromosome. Lastly, the selection phase is defined by a roulette based operation. Experiments were performed to investigate the behavior of the proposed method with example instances of Lingo 8.0 and the results were compared with this solver. The genetic algorithm was able to obtain better solutions only in 2 of 9 instances.

LocalSolver [14] is local search based heuristic commercial solver that uses local search to optimize linear and nonlinear binary problems. By default, LocalSolver performs a descent method as search strategy using autonomous movements. A Simulated Annealing based algorithm is also included. Both of these strategies are implemented in multithreading. Initial solution is found by a basic randomized greedy algorithm. *Autonomous movements*, which are k-flips movements, are used, generating movements similar to Ejection Chains. These movements preserve the feasibility of a solution. In its latest version, LocalSolver reaches better solutions than Gurobi and CPLEX solvers in some MIPLIB 2010 instances classified as hard. Both local search and the constructive algorithm are only superficially described by the authors, probably because of the commercial nature of the product.

Vassilev et al. [15] presents a hybrid heuristic algorithm for Mixed Integer Programming were each iteration has polynomial-time computing complexity. This algorithm searches for feasible integer directions and uses a linear solver for the continuous part. Three solutions are provided and the best of them is chosen. The method proceeds, iteratively generating subproblems which the feasible region is defined by all problem constraints satisfied at the current iteration. In these subproblems the objective function is one of the constraints which are not been satisfied yet. When a feasible solution was found, the objective function of the original problem is inserted on the next subproblems. Any solution found from this stage leads to a gradual improvement. Tests compare two versions of the method, one that uses feasible integer directions with one non-zero component and other that uses two of this components. The first variant shows solutions with better quality and execution time, besides occupying a smaller portion of memory.

3   The BP Local Search Solver

Before proceeding to a detailed description of our solver it is important to comment about the diversity of constraint types which can appear in BP problems and how it determines the hardness of finding a first feasible solution. Some quite common constraint types are:

Set Covering:	$\sum x_i \geq 1$, x_i is binary
Set Packing:	$\sum x_i \leq 1$, x_i is binary
Set Partition:	$\sum x_i = 1$, x_i is binary

While for some BPs a feasible solution is trivial, e.g. Set Covering or Set Packing, different constraints can significantly complicate this initial step. The satisfaction of just one constraint can be a NP-Complete problem if it represents, for instance the Number Partitioning Problem. Problems where only few, *hidden*, subsets of all possible incidence vectors are feasible tend to be much harder. Set Partitioning problems are typical examples of this type.

Our solver discovers and explores relationships between variables both in the constructive phase and in the local search phase by means of conflict graphs, which will be explained in the next subsection.

3.1   Conflict Graphs

A fundamental information from BP used by Linear Programming (LP) based solvers to generate cuts and to strengthen the LP relaxation is the conflict graph [16]. In ou work the conflict graph is always used in the primal search space, both in our constructive approach and in the local search approach.

We construct a conflict graph by detecting pairs of variables which cannot be activated at the same time. Since the construction of a full conflict graph may require the execution expensive techniques such as probing [17], we opted for a simpler procedure: conflicts are detect by processing each constraint individually, checking for pairs of variables in this constraint whose activation cannot occur at the same time without violating it.

To illustrate relationships between variables and conflict graph consider the binary program $\mathcal{P}$:

$$min.:$$
$$10x_1 + 12x_2 + 4x_3 + 7x_4 + 5x_5$$
$$s.t.:$$

$$x_1 + x_2 \leq 1 \tag{4}$$
$$x_1 + x_3 + x_5 = 1 \tag{5}$$
$$x_2 + x_4 \geq 1 \tag{6}$$
$$x_2 + x_4 + x_5 \leq 1 \tag{7}$$
$$x_1, x_2, x_3, x_4, x_5 \in \{0, 1\}$$

In $\mathcal{P}$, Set Partitioning (5) and Set Packing constraints (4 and 7) are the constraints which provide obvious sets of conflicting variables. More generally, Generalized Upper Bounds (GUB) constraints are rich sources of conflicts. The conflict graph for $\mathcal{P}$ can be seen in Figure 1. Connected nodes represent conflicting variables. Besides conflicts this graph also shows a different relationship between variables: dashed lines are drawn in groups of variables where at least one variable must be activated.

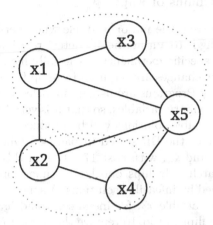

Fig. 1. Conflict Graph for $\mathcal{P}$

3.2   Constructive Approach

An initial solution is built by solving a subproblem considering only the structure of set partition, packing and covering constraints. Solving this subproblem corresponds to find an independent set in the conflict graph, i.e. finding a set of variables that have no conflict with each other.

Finding and independent set corresponds to finding a clique in the complimentary graph. Thus, the subproblem is modeled as a graph which is complementary to the original conflict graph. The weight of each vertex is related to the number of Packing and Covering constraints that it satisfies when activated. A Tabu Search based algorithm [18] then searches for cliques with weight above a threshold given as input. In this case, the threshold is the number of set partition and set covering constraints contained in the problem. When the algorithm finishes, a solution is created for the original problem, activating only the variables returned by Tabu Search.

This phase was developed to obtain an initial set of variables that can be activated without generating infeasibilities among them. In the case of instances which contain only these three types of constraints, the result of the constructive phase is feasible solution for the original problem. Otherwise, the returned solution can be infeasible, so that infeasible constraints are relaxed and sent to be fixed in the local search phase. In our experiments we observed that even when the BP has many other constraints, the initial satisfaction of these three constraint types speeds up a lot the search process, since remaining constraints are quite often easier to satisfy.

3.3   Local Search : Chains of Flips

As the authors of [14] noted, the major obstacle when performing local search with binary programming is to automatically detect relationships between decision variables. Problem specific metaheuristics usually have a compact solution representation where few changes are required to jump from one solution to another. In contrast, Binary Programs are usually modeled a much larger number of decision and related auxiliary variables, so that it is very likely that by flipping one bit at time only unfeasible solutions will be produced.

As an example, consider the BP stated in page 48. One feasible solution is to activate variables x_1 and x_4, with cost 17. Once in this feasible solution, a method whose local search only flips one bit at time would be trapped in a local optimum surrounded by infeasible solutions. A smarter solver, when trying to flip the then inactive variable x_2, for instance, would have to automatically detect that x_1 should be flipped too, to remove the conflict caused by constraint 4, and that another variable should now flipped to satisfy constraint 5, say x_3, which would be infeasible. Thus, when trying to flip the then inactive variable x_2 the solver would detect a *chain* of movements: $x_2 \rightarrow x1 \rightarrow x3$, where every subsequent move would fix an infeasibility caused by a previous move. This specific chain would produce a better solution with cost 16.

A fast algorithm to search for these chains of movements is the key component to every local search based method for BP. In [14], even though authors comment about the importance of this step, no details are given about how these chains are generated, probably because the referred paper describes a commercial product.

Algorithm 1.1 describes our implementation of an algorithm to detect a chain of movements which lead from one feasible solution to another. The algorithm performs a backtracking with limited depth $\overline{d}$ and limited breadth $\overline{f}$. At each recursion a set $\hat{J}$ of variables are flipped: the current variable j and all conflicting variables, if j will become an active variable. These variables are put in a *frozen* state (set S) in this and in deeper levels of the recursion. Subsequent variables to be flipped are chosen from a set $\tilde{J}$ of variables. To fix new infeasibilities, only variables which appear on the constraints set C can help. Candidate variables $\tilde{j}$ are evaluated with respect to how many conflicts it decreases in constraints of C, this evaluation is stored in $e_{\tilde{j}}$. The most promising variables flips are further evaluated recursively in lines 21 to 26 and if the final effect is positive then the recommended chain of movements J^* is augmented. As one can observe, a smart computation of $\tilde{j}_j$ is a key point to the success of the method, since large values of $\overline{d}$ and $\overline{f}$ would result in prohibitive computing times. In this sense, we observed that besides prioritizing variables which decrease the largest amount of infeasibilities we also should include in this evaluation a larger priority to variables which decrease infeasibilities in constraints with less options to resolve these infeasibilities.

Algorithm 1.1. chainFlip

Input :

    x: current solution;

    j: variable to be flipped;

    J: variables already flipped;

    C: constraints to check;

    S: frozen variables;

    d: current depth;

    $\bar{d}$: maximum depth;

    $\bar{f}$: maximum number of flips per recursive call;

Output:

    (z^*, J^*): cost and variables of the best chain found

1 **if** $d \geq \bar{d}$ **then** return

2 $\hat{J} = \{j\}$;

3 **if** $x_j = 0$ **then**

4 | $S \leftarrow S \cup \{j'\} : conflict(j, j')$;

5 | $\hat{J} \leftarrow \hat{J} \cup \{j'\} : conflict(j, j') \wedge x_{j'} = 1$;

6 **end if**

7 $x' = x$;

8 **for** $j' \in \hat{J}$ **do**

9 | $x'_j = 1 - x'_j$

10 **end for**

11 $J \leftarrow J \cup \hat{J}$;

12 $z^* \leftarrow f(x')$;

13 $J^* \leftarrow J$;

14 $C \leftarrow C \cup \{i\} : i$ is a constraint where one or more variables of $\hat{J}$ appear;

15 $\tilde{J} \leftarrow$ all j which appear in some constraint of C and is not in S;

16 $e_{\tilde{j}} = 0, \; \forall \tilde{j} \in \tilde{J}$;

17 **for** $\tilde{j} \in \tilde{J}$ **do**

18 | compute the impact $e_{\tilde{j}}$ of flipping $\tilde{j}$ considering constraints C;

19 **end for**

20 **for** $k = 1$ *to* $min(\bar{f}, |\tilde{J}|)$ **do**

21 | $\tilde{j} \leftarrow$ the $k-$th element from $\tilde{J}$ with best $e_{\tilde{j}}$;

22 | $(z', J') \leftarrow$ chainFlip$(x', \tilde{j}, J, C, S, d+1, \bar{d}, \bar{f})$;

23 | **if** $z' < z^*$ **then**

24 | | $z^* \leftarrow z'$;

25 | | $J^* \leftarrow J$;

26 | **end if**

27 **end for**

28 return (z^*, J^*);

The current implementation of local search performs iterated calls to chainFlip made from different, randomly selected variables. If the solution is still unfeasible, the search concentrates in variables which appear in constraints which are still not satisfied. In this case, we first randomly select one of the unfeasible constraints and then randomly select one of its variables. Movements are accepted according to a RNA (Random Non Ascendent) rule. As it can be seen in the next section, this simple approach, combined with our constructive algorithm was enough to produce very encouraging results.

4   Computational Experiments

Our code was written in C++ using the open source COIN-OR libraries to read instances. The code was compiled on GCC/G++ version 4.6.3. We ran all the experiments on an Intel Core i7-3770 ® 3.4GHz computer with 16Gb of RAM running the openSUSE Linux 12.3 operating system.

Table 1. Details of instances

Instance	Rows	Cols	Objective	COV	PAC	PAR
acc-tight5	3052	1339	0.00	11	288	244
air04	823	8904	56137.00	0	0	823
bab5	4964	21600	-106412.00	0	88	21
bley_xl1	175620	5831	190.00	14	5133	169
bnatt350	4923	3150	0.00	183	0	0
cov1075	637	120	20.00	252	0	0
eil33.2	32	4516	934.01	0	0	32
eilB101	100	2818	1216.92	0	0	100
ex9	40962	10404	81.00	0	0	162
iis-100-0-cov	3831	100	29.00	3831	0	0
iis-bupa-cov	4803	345	36.00	4803	0	0
iis-pima-cov	7201	768	33.00	7201	0	0
m100n500k4r1	100	500	-25.00	0	100	0
macrophage	3164	2260	374.00	609	0	0
mine-166-5	8429	830	-5.66E+08	0	0	0
mine-90-10	6270	900	-7.84E+08	0	0	0
mspp16	561657	29280	363.00	15	1695	31
n3div36	4484	22120	130800.00	2	4424	0
n3seq24	6044	119856	52200.00	120	4484	0
neos-1109824	28979	1520	378.00	0	1520	23
neos-1337307	5687	2840	-202319.00	0	0	126
neos18	11402	3312	16.00	2809	0	2262
neos-849702	1041	1737	0.00	0	540	270
netdiversion	119589	129180	242.00	103	49799	1
ns1688347	4191	2685	27.00	0	382	88
opm2-z7-s2	31798	2023	-10280.00	0	0	0
reblock67	2523	670	-3.46E+07	0	0	0
rmine6	7078	1096	-457.19	0	0	0
sp98ic	825	10894	4.49E+08	6	627	0
tanglegram1	68342	34759	5182.00	7843	0	0
tanglegram2	8980	4714	443.00	2160	0	0
vpphard	47280	51471	5.00	0	0	320

Table 2. Production of feasible solutions in 60 and 300 seconds

Instance	60 seconds			300 seconds		
	GLPK	CBC	BPLS	GLPK	CBC	BPLS
acc-tight5						
air04	✓	✓	✓	✓	✓	✓
bab5				✓		
bley_xl1						
bnatt350						
cov1075	✓	✓	✓	✓	✓	✓
eil33-2	✓	✓	✓	✓	✓	✓
eilB101	✓	✓	✓	✓	✓	✓
ex9						
iis-100-0-cov	✓	✓	✓	✓	✓	✓
iis-bupa-cov	✓	✓	✓	✓	✓	✓
iis-pima-cov	✓	✓	✓	✓	✓	✓
m100n500k4r1	✓	✓	✓	✓	✓	✓
macrophage	✓	✓	✓	✓	✓	✓
mine-166-5	✓	✓	✓	✓	✓	✓
mine-90-10			✓		✓	✓
mspp16						
n3div36		✓	✓	✓	✓	✓
n3seq24			✓		✓	✓
neos-1109824	✓	✓		✓	✓	✓
neos-1337307	✓	✓		✓	✓	
neos18	✓	✓	✓	✓	✓	✓
neos-849702						
netdiversion						
ns1688347						
opm2-z7-s2	✓	✓	✓	✓	✓	✓
reblock67		✓	✓	✓	✓	✓
rmine6	✓	✓	✓	✓	✓	✓
sp98ic	✓	✓	✓	✓	✓	✓
tanglegram1		✓				✓
tanglegram2	✓	✓	✓	✓	✓	✓
vpphard					✓	
Total	**17**	**19**	**20**	**19**	**23**	**21**

Computational experiments were made using all 32 binary problems of MI-PLIB 2010 benchmark set [19] which have a feasible solution. Since its introduction in 1992, the MIPLIB became a standard library of tests used to compare the performance of integer programming solvers. It contains a collection of real problems, most of them based on industrial applications. The details of the used problems can be seen in Table 1. Columns Rows and Cols indicate the number of constraints and decision variables of the problems, respectively. Column Objective presents the optimal objective value for each instance. The remaining columns COV, PAC and PAR indicates the number of set covering, set packing and set partition constraints for each instance, respectively.

These experiments compare our approach with two of the best open source integer programming solvers: CBC[1] and GLPK[2]. Table 2 shows the obtained results with execution time limit set to 60 and 300 seconds. In this table, columns GLPK and CBC indicate, respectively, tests performed using GLPK and CBC solvers with default parameters. The Last column, BPLS, corresponds to results obtained by our approach. In both of these columns, a check mark is used to indicate the method has found a feasible solution for a instance in the restricted time limit.

Results show that our approach was able to find feasible solutions to a greater number of instances in a 60 seconds timeout when comparing with CBC and GLPK. Relaxing this time limit to 300 seconds, all methods were able to find more feasible solutions. While our approach found feasible solutions for 21 instances, GLPK and CBC found 19 and 23 feasible solutions, respectively.

5   Conclusions

In this work we proposed and evaluated computationally a hybrid, local search based solver to search for feasible solutions for Binary Programming problems. Computational experiments performed in the MIPLIB 2010 instance set showed that our approach is more reliable to find feasible solutions in very restricted amounts of time than two of the best open source integer programming solvers available: CBC and GLPK.

This feature is fundamental for those interested in the application of Binary Programming where time is a limiting factor. It is also worth to note that the production of the first feasible solution can also speed up the production of high quality solutions: once a feasible solution is available methods like RINS or Local Branching can me immediately applied to improve the incumbent solution.

References

1. Dantzig, G., Fulkerson, R., Johnson, S.: Solution of a Large-Scale Traveling-Salesman Problem. Journal of the Operations Research Society of America 2(4), 393–410 (1954)
2. Pritsker, A.A.B., Watters, L.J., Wolfe, P.M.: Multiproject Scheduling with Limited Resources: A Zero-One Programming Approach. Management Science 16(1), 93–108 (1969)
3. Lancia, G.: Integer programming models for computational biology problems. Journal of Computer Science and Technology 19(1), 60–77 (2004)
4. Johnson, E., Nemhauser, G., Savelsbergh, W.: Progress in Linear Programming-Based Algorithms for Integer Programming: An Exposition. INFORMS Journal on Computing 12 (2000)
5. Linderoth, J.T., Ralphs, T.K.: Noncommercial Software for Mixed-Integer Linear Programming. In: Karlof, J.K. (ed.) Integer Programming: Theory and Practice, pp. 253–303. CRC Press Operations Research Series (2005)

---

[1] https://projects.coin-or.org/Cbc
[2] http://www.gnu.org/software/glpk/

6. Land, A.H., Doig, A.G.: An Automatic Method of Solving Discrete Programming Problems. Econometrica 28(3), 497–520 (1960)
7. Mittelman, H.: Feasibility benchmark (February 2014), http://plato.asu.edu/ftp/feas_bench.html
8. Lougee-Heimer, R.: The Common Optimization INterface for Operations Research: Promoting open-source software in the operations research community. IBM Journal of Research and Development 47(1), 57–66 (2003)
9. Forrest, J., Lougee-Heimer, R.: CBC User Guide. INFORMS Tutorials in Operations Research, 257–277 (2005)
10. Fischetti, M., Glover, F., Lodi, A.: The feasibility pump. Mathematical Programming 104, 2005 (2005)
11. Fischetti, M., Lodi, A.: Local branching. Mathematical Programming 98(1-3), 23–47 (2003)
12. Danna, E., Rothberg, E., Pape, C.L.: Exploring relaxation induced neighborhoods to improve mip solutions. Mathematical Programming 102(1), 71–90 (2005)
13. Huy, P.N.A., San, C.T.B., Triantaphyllou, E.: Solving integer programming problems using genetic algorithms. In: ICEIC: International Conference on Electronics, Informations and Commnications, pp. 400–404 (2004)
14. Benoist, T., Estellon, B., Gardi, F., Megel, R., Nouioua, K.: Localsolver 1.x: a black-box local-search solver for 0-1 programming. 4OR 9(3), 299–316 (2011)
15. Vassilev, V., Genova, K.: An algorithm of internal feasible directions for linear integer programming. European Journal of Operational Research 52(2), 203–214 (1991)
16. Atamtürk, A., Nemhauser, G.L., Savelsbergh, M.W.P.: Conflict graphs in solving integer programming problems. European Journal of Operational Research 121, 40–55 (2000)
17. Borndorfer, R.: Aspects of Set Packing, Partitioning, and Covering. PhD thesis (1998)
18. Wu, Q., Hao, J.K., Glover, F.: Multi-neighborhood tabu search for the maximum weight clique problem. Annals of Operations Research 196(1), 611–634 (2012)
19. Koch, T., Achterberg, T., Andersen, E., Bastert, O., Berthold, T., Bixby, R., Danna, E., Gamrath, G., Gleixner, A., Heinz, S., Lodi, A., Mittelmann, H., Ralphs, T., Salvagnin, D., Steffy, D., Wolter, K.: Miplib 2010. Mathematical Programming Computation 3(2), 103–163 (2011)

A Partition-Based Heuristic for the Steiner Tree Problem in Large Graphs*

Markus Leitner[1], Ivana Ljubić[1], Martin Luipersbeck[2], and Max Resch[2]

[1] Department of Statistics and Operations Research, University of Vienna, Vienna,
Austria
{markus.leitner,ivana.ljubic}@univie.ac.at
[2] Vienna University of Technology, Austria
{martin.luipersbeck,max.resch}@alumni.tuwien.ac.at

Abstract. This paper deals with a new heuristic for the Steiner tree problem (STP) in graphs which aims for the efficient construction of approximate solutions in very large graphs. The algorithm is based on a partitioning approach in which instances are divided into several subinstances that are small enough to be solved to optimality. A heuristic solution of the complete instance can then be constructed through the combination of the subinstances' solutions. To this end, a new STP-specific partitioning scheme based on the concept of Voronoi diagrams is introduced. This partitioning scheme is then combined with state-of-the-art exact and heuristic methods for the STP. The implemented algorithms are also embedded into a memetic algorithm, which incorporates reduction tests, an algorithm for solution recombination and a variable neighborhood descent that uses best-performing neighborhood structures from the literature. All implemented algorithms are evaluated using previously existing benchmark instances and by using a set of new very large-scale real-world instances. The results show that our approach yields good quality solutions within relatively short time.

1 Introduction

The Steiner tree problem (STP) in graphs is a fundamental $\mathcal{NP}$-hard combinatorial optimization problem with numerous applications, e.g., in telecommunication network design or computational biology. In the STP we are given an undirected graph $G = (V, E)$ whose node set V is the disjoint partition of *terminal nodes* T, $\emptyset \neq T \subset V$, and *potential Steiner nodes* $V \setminus T$ as well as a cost function $c : E \to \mathbb{Q}_+$ assigning a nonnegative value to each edge. The goal is to find a subgraph $S = (V_S, E_S)$, $V_S \subseteq V$, $E_S \subseteq E$, of G spanning all terminals, i.e., $T \subseteq V_S$, of minimum cost $c(E_S) = \sum_{e \in E_S} c_e$. The STP in graphs has received significant attention from the scientific community in the last decades. Several integer linear programming (ILP) formulations together with corresponding solution methods have been developed [1] and the lower bounds arising from their

* This work is supported by the Austrian Science Fund (FWF) under grants I892-N23 and P26755-N26.

M.J. Blesa, C. Blum, and S. Voß (Eds.): HM 2014, LNCS 8457, pp. 56–70, 2014.
© Springer International Publishing Switzerland 2014

linear programming relaxations as well as from Lagrangian relaxation or dual ascent have been studied, see, e.g., [2,3,4,5]. The current state-of-the-art approach
for solving instances of the STP to proven optimality has been proposed by
Polzin [4] and Daneshmand [6]. Their approach incorporates several algorithmic
techniques such as reduction tests [7], dual ascent, and construction of heuristic solutions, within a branch-and-bound framework. Aside from exact methods
numerous metaheuristic approaches have been applied to the STP to compute
good solutions within shorter time, see e.g., [8,9,10].

In this work, we propose a new *matheuristic* and a memetic algorithm for the
STP. Our approaches, that will be detailed in the following, aim to effectively
solve very large-scale instances through the combination of graph partitioning
and state-of-the-art exact and heuristic methods for the STP. Note, that most
components of the algorithms discussed in the following are described in more
detail in the Master's thesis of the third author of this work [11].

2 Partition-Based Construction Heuristic

This section details the partition-based construction heuristic (PCH), which applies graph partitioning as a means to find a heuristic problem decomposition.
A given STP instance is heuristically divided into a set of smaller subinstances,
which are solved separately and whose solutions are combined into a feasible
solution to the original instance. In the past this concept has been applied to
compute good feasible solutions to large-scale instances of other $\mathcal{NP}$-hard problems which require too much computational effort for current exact methods,
see, e.g., [12]. A general framework which follows a similar principle has been
proposed by Taillard and Voß [13]. Figure 1 visualizes the general framework and
its four main steps (*partition, decompose, solve,* and *repair*) that are described in
the following subsections, while Figure 2 depicts each stage of the process when
applied to a simple problem instance. Note that we assume that preprocessing
in the form of reduction tests has been applied prior to the PCH.

2.1 Step 1: Partition

The goal of the first step is to compute a partition of the instance graph, which
is used later on to decompose the given STP instance into subinstances. As
indicated in Figure 1, given the preferred number of partitions k and a partition
imbalance parameter d, $1 \leq d \leq k$, a partitioning algorithm $\mathcal{A}_P$ is applied
to divide the given graph G into k subsets each containing no more than $d \cdot
\frac{|V|}{k}$ nodes. The parameters k and d enable a trade-off between solution quality
and runtime, i.e., a high number of small subinstances is solved easily, but the
resulting solution quality may in turn be low.

For heuristic problem decomposition to yield good quality solutions, the independence of the subinstances plays an important role. Given an STP instance
and a set of subinstances thereof, we consider the subinstances as independent if
they can be solved separately, such that the union of their solutions corresponds

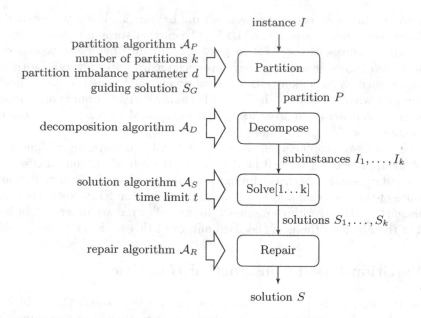

Fig. 1. A partition-based procedure for heuristic solution construction

to the optimal solution of the original instance. In the context of the $\mathcal{NP}$-hard STP, completely independent subinstances do not exist in general (although in some special cases an exact decomposition is feasible [4]).

Thus we propose two heuristic graph-structure-based measures which describe the independence between potential subinstances: the *edge-cut* between subgraphs and the *distance between terminals*. The first measure is aimed at graphs of varying density, while the second measure focuses on instances that contain terminal clusters.

Furthermore, we introduce the concept of the *guiding solution* S_G to enhance our partition-based decomposition algorithms. A guiding solution is a heuristic solution to the given instance as a whole, potentially computed by a fast construction heuristic able to efficiently process large-scale instances.

Both algorithms proposed in this article aim to construct a partitioning that splits S_G into k subtrees. The goal is to group terminals together into the same subset, if they are connected by a short path in S_G. If S_G is already a good approximation of the optimal solution's structure, the subinstances' solutions are also likely to be similar to parts of the optimal solution.

Edge-based Partitioning (Eb). The objective of this algorithm is to find a partitioning that minimizes the edge-cut of the instance graph. Given a graph $G = (V, E)$ with node weights w_i, $i \in V$, edge weights c_{ij}, $\{i, j\} \in E$, an integer $k > 1$ and a partition imbalance parameter d, the goal is to find a *balanced partition* of V into k disjoint subsets $V_1, \ldots, V_k$ such that $\bigcup_{i=1}^{k} V_i = V$ and such that the weight of the edges between different subsets is minimized. Thereby, by *balanced*

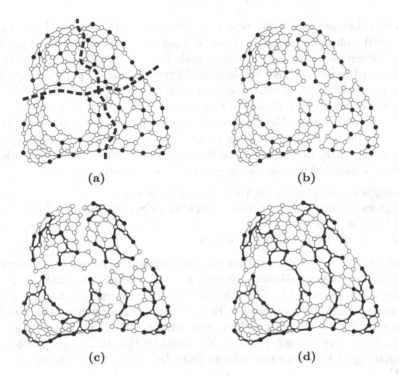

Fig. 2. PCH solution construction. (a) Graph after computing a partition (partitions marked by the dashed lines and terminals in black). (b) Subinstances obtained from decomposition. (c) Solutions of subinstances (bold edges indicate computed solutions). (d) Solution to original instance after repair.

we mean that $\sum_{i \in V_i} w_i \leq \frac{d}{k} \cdot \sum_{i \in V} w_i$ holds for each subset V_i, $1 \leq i \leq k$. For the purpose of problem decomposition for the STP, we assume that a heuristic partition suffices. In our implementation this task is handled by an efficient heuristic implemented in the publicly available partitioning framework METIS [14], which performs in linear time with respect to the number of nodes.

In our experiments we have considered two *weighting schemes* with either uniform edge weights $c'_e = 1$, $\forall e \in E$, (minimizing the size of the edge-cut) or transformed original weights $c'_e = c_{max} - c_e$, $\forall e \in E$, with $c_{max} = \max_{e \in E} c_e$ (minimizing the weight of the edge-cut). For both weighting schemes we incorporate heuristic information provided by the guiding solution S_G through weight scaling, where the weights c'_e of edges $e \in S_G$ are scaled by a certain priority factor l to become heavier than regular edges. The goal is to make these edges less likely to be included into an edge-cut, since partition subsets should be computed so that they decompose the guiding solution into subtrees. For our experiments we have chosen $l = 2$.

Voronoi-based Partitioning (Vb). Voronoi diagrams in graphs have been successfully applied in the STP literature to design efficient local search and reduction techniques [4,10]. We now propose a new partitioning scheme for the STP derived from the concept of Voronoi diagrams. In the context of the STP, a Voronoi-diagram assigns each Steiner node to its nearest terminal, while ties are broken arbitrarily. Thus the Voronoi diagram defines a partitioning of the instance graph, where each subset contains exactly one terminal, as well as all its closest Steiner nodes. Let P denote this initial partitioning, and let $p \in P$ and $p' \in P$ be disjoint subsets. Furthermore, let $d[p, p']$ denote the minimum distance between any pair of terminals t_1 and t_2, where $t_1 \in p$ and $t_2 \in p'$. Subsequently, P is iteratively coarsened until it is sufficiently close to the specified partitioning parameters k and d. The coarsening process is performed as follows:

1. Choose a subset $p \in P$ such that $|p| \leq |p'|$, $\forall p' \in P$.
2. Merge p with an adjacent subset p', such that $d[p, p'] \leq d[p, p'']$, $\forall p'' \in P \backslash \{p\}$, and $|p| + |p'| \leq |V| \cdot \frac{d}{k}$.
3. If $|P| > k$ and there exists p', go to 1.

Always choosing the subset with minimum cardinality ensures that the created partitioning is roughly balanced, since two large subsets can only be merged if no smaller subsets exist. Note that this procedure ensures that partitions adhere to the imbalance parameter d. However, it may happen that the algorithm terminates while more than k subsets remain, since no p' may exist such that the balance property is not violated. We consider this to be acceptable, since our primary goal is to prevent subsets from becoming too large to be solved efficiently.

To incorporate information provided by a guiding solution S_G, in Step 2 of the procedure, subsets are only merged if there exists a path between them in S_G. Thereby, two subsets are only contracted if a path connects them directly in S_G, i.e., without passing through another subset first.

2.2   Step 2: Decompose

During the decomposition step a set of subinstances $I_1, \ldots, I_k$ is constructed from the original instance $I = (G, T)$, where $G = (V, E)$ is the instance graph and $T \subset V$ is the set of terminals. The decomposition is performed based on the partition P of G computed in the previous step. We consider the following decomposition strategy which turned out to perform best among different alternatives considered, see [11] for more details. Given the partition P, the edges E_P corresponding to the edge-cut defined by P are removed from E. Thus G is split into a set of subgraphs $G_1, \ldots, G_k$. From each subgraph $G_i = (V_i, E_i)$, a subinstance $I_i = (G_i, T_i)$ is constructed, where $T_i = V_i \cap T$. Note that this method implies that the union of all solutions S_i of I_i will only form a partial solution in I, since all subinstances are disjunct. The remaining edges have to be computed in Step 4 (repair).

2.3   Step 3: Solve

The algorithm used for the exact solution of the STP which is also used for solving subinstances is based on a branch-and-cut (B&C) approach similar to the one proposed by Koch and Martin [1], see also [5]. For each arc $(i, j) \in A = \{(i,j) \mid \{i,j\} \in E\}$, an arc variable x_{ij} denotes membership of the corresponding arc to the Steiner tree $(x_{ij} = 1)$ or not $(x_{ij} = 0)$. Similarly, additional node variables y_i for $i \in V \setminus T$ denote if i is spanned by the Steiner tree $(y_i = 1)$ or not $(y_i = 0)$. An arbitrary terminal is chosen as root node r. For brevity, we use the following notations: Given a set $W \subset V$, we define $\delta^+(W) = \{(i,j) \in A \mid i \in W \wedge j \in V \setminus W\}$ as the set of all arcs with tail inside W and the head in its complement. Conversely, $\delta^-(W)$ denotes the set of arcs pointing into W from its complement set. For short, if W contains only a single element v, we write $\delta^+(\{v\})$ as $\delta^+(v)$ and $\delta^-(\{v\})$ as $\delta^-(v)$, respectively.

$$\text{(EDCF)} \quad \min \left\{ \sum_{(i,j)\in A} c_{ij}x_{ij} \mid (x,y) \in \{0,1\}^{|A|+|V|-|T|} \right.$$

$$x(\delta^-(i)) = 1, \ \forall i \in T \setminus \{r\}, \quad x(\delta^-(i)) = y_i, \ \forall i \in V \setminus T \tag{1}$$

$$\left. x(\delta^-(W)) \geq 1, \ \forall \ W \subset V, \ r \notin W, \ W \cap T \neq \emptyset \right\} \tag{2}$$

The objective function minimizes the weight of the selected arcs. Degree constraints (1) ensure that each terminal except the root and all Steiner nodes that are part of the solution have in-degree exactly one. Constraints (2) are directed cut constraints that ensure that there is a directed path between the root and any other terminal node.

The following inequalities are additionally used to initialize the branch-and-cut procedure:

$$x(\delta^+(i)) \geq y_i \qquad\qquad \forall i \in V \setminus T \tag{3}$$

$$x_{ij} + x_{ji} \leq y_i \qquad\qquad \forall (i,j) \in A, i \in V \setminus T \tag{4}$$

Constraints (3) ensure that Steiner nodes that are part of the solution have at least one outgoing arc (they were referred to as "flow-balance" constraints in the literature). Constraints (4) express that each arc in the solution tree can only be oriented in one way. We also add root in- and out-degree constraints: $x(\delta^+(r)) \geq 1$ and $x(\delta^-(r)) = 0$ (notice that one can alternatively remove root-incoming arcs from the input graph).

Since formulation (EDCF) contains an exponential number of directed cut constraints (2) we implemented a branch-and-cut algorithm. The branch-and-cut is initialized with all compact constraints and with a set of cut constraints obtained through Wong's dual ascent algorithm [5]. Thus, the linear programming (LP) bound obtained from our initial model is at least as good as the one obtained from dual ascent and hence a significant reduction of the runtime can typically be observed, cf. [2,4]. We also initialize the upper bound using the feasible solution obtained from dual ascent. Further cut constraints are separated using the *push-relabel maximum flow* algorithm [15] and we also used nested cuts, back cuts and creep-flow to improve the number and strength of separated inequalities per call of the separation routine, see [1] for details. Our branch-and-cut incorporates a *primal heuristic* which is called after each cutting-plane iteration, in order to compute tight upper bounds that enable effective pruning of nodes of the search tree. We apply the improved implementation of the well-known shortest path heuristic (SPH) [16], combined with a pruning step (MST-Prune) as proposed in [17], which achieves a much better average-case runtime than the classic implementation. MST-Prune can potentially enhance the solution constructed by SPH by computing a minimum spanning tree on the set of nodes contained in the solution and recursively removing Steiner nodes of degree one. SPH followed by MST-Prune is applied to the original undirected graph with adapted edge weights $c'_{ij} = c_{ij} \cdot (1 - \max(\tilde{x}_{ij}, \tilde{x}_{ji})), \forall \{i, j\} \in E$, which are computed from the current LP solution $(\tilde{x}, \tilde{y})$.

2.4 Step 4: Repair

As mentioned above, solving the subinstances generated in the decomposition step results in a set of disconnected partial solutions. Thus a repair step is necessary to extend these into a feasible solution for the original instance. Given a set of solutions S_i, $1 \leq i \leq k$, to subinstances I_i, the goal is to identify a set of edges E' such that $\bigcup_{1 \leq i \leq k} S_i \cup E'$ is connected.

In our implementation the edges E' are computed through the construction of an appropriate *auxiliary STP instance* to which either a *heuristic* or *exact* algorithm can be applied. The auxiliary instance is obtained from the original graph in which partial solutions S_i, $1 \leq i \leq k$ are shrunk into "super-nodes" (more precisely, "super-terminals"), self-loops are deleted, and among parallel edges only the cheapest ones are kept. Subsequently, (1) for the heuristic repair: SPH is performed followed by MST-Prune with a randomly selected terminal as root, and (2) for the exact repair we use the branch-and-cut algorithm detailed above.

3   Partition-Based Memetic Algorithm

In this section, we present a memetic algorithm (to which we refer to as *MPCH*) in which PCH is applied in combination with several other problem-specific algorithms. The objective is to exploit synergy effects arising from the interaction between multiple algorithmic components.

Within MPCH, PCH is always supplied with already available heuristic information in the form of a guiding solution and therefore does not construct solutions from scratch. Given a population of solutions, PCH can be interpreted as a specialized mutation operator, which introduces new information and potentially enhances a given solution. Its application is also similar to the iterative improvement provided by a local search procedure. In the proposed algorithm PCH is not only applied to a solution once, but several times. The solution produced in the previous iteration is subsequently used as a guiding solution in the next step. This procedure is allowed to continue until no improving solution has been found for a specified number of iterations. In the end, the best found solution replaces the original guiding solution in the population.

Data: Instance $I = (G, T, c)$, population size $popmax$, maximum number of
    generations g, number of iterations without improvement n
Result: The best found solution S^* and an associated lower bound lb.

$pop \leftarrow \emptyset$
$lb \leftarrow 0$

for $popmax$ *individuals* **do**   // Initialize population
$\quad | \quad (lb', G_A, \tilde{c}) \leftarrow \text{DA}(I)$
$\quad | \quad I' \leftarrow (G_A, T, c)$
$\quad | \quad S \leftarrow \text{SPH}(I')$
$\quad | \quad S \leftarrow \text{VND}(I, S)$

$\quad | \quad \text{insertInPopulation}(pop, S)$
$\quad | \quad lb \leftarrow \max(lb, lb')$
$\quad | \quad \text{reduce}(I, \tilde{c}, lb, \text{obj}(\text{best}(pop)))$
end

for g *generations* **do**   // Generation step
$\quad | \quad$ **foreach** $S \in pop$ **do**
$\quad | \quad \quad | \quad S' \leftarrow S$
$\quad | \quad \quad | \quad$ **repeat**
$\quad | \quad \quad | \quad \quad | \quad S \leftarrow \text{PCH}(I, S)$
$\quad | \quad \quad | \quad \quad | \quad S \leftarrow \text{VND}(I, S)$
$\quad | \quad \quad | \quad \quad | \quad S' \leftarrow min_{obj}(S', S)$
$\quad | \quad \quad | \quad$ **until** n *iterations without improvement*
$\quad | \quad \quad | \quad \text{replaceInPopulation}(pop, S, S')$
$\quad | \quad$ **end**
$\quad | \quad pop \leftarrow \text{recombination}(pop)$
end

$S^* \leftarrow \text{best}(pop)$

Algorithm 1. Partition-based Memetic Algorithm

Algorithm 1 shows the structure of MPCH whose components will be detailed
below. Basic population-based parameters are the population size $popmax$ and
the maximum number of generations g. The parameter n restricts the number
of PCH applications without improvement. The whole procedure returns the
best found solution S^* and an associated lower bound lb as an estimate of the
solution's quality. In the following, let $\text{best}(pop)$ return the best solution of the
population pop with respect to the objective value, and let $\text{obj}(S)$ denote the
objective value of a solution S.

Generation of an initial population. In each iteration, the dual ascent algorithm
(DA) [5] is executed with an arbitrary terminal as root node. The result is a
saturation graph G_A, a lower bound lb' and the reduced costs $\tilde{c}$. Subsequently, the
shortest path heuristic (SPH) is applied to G_A to construct a feasible solution.
The result is improved through the application of variable neighborhood descent
(VND), see below. The resulting solution S is inserted into the population pop.

Finally, the currently best lower bound lb and upper bound (i.e., the objective value of the best solution in pop), as well as the reduced costs $\tilde{c}$ from the current iteration are used to apply bound-based reduction tests to the instance I. After the population initialization, one obtains a population pop of (diverse) feasible solutions together with a lower bound lb and a reduced instance graph G.

Generation step. In the main phase, the population is evolved for a fixed number of generations, each of which consisting of two steps: individual improvement and solution recombination. In the improvement step, PCH followed by VND is applied in a multi-start fashion to each solution $S \in pop$. For the first iteration of the multi-start procedure, S is used as a guiding solution for PCH. The guiding solution for each subsequent iteration is the solution from the previous iteration. The multi-start procedure continues until no improving solution has been found within n consecutive iterations. After the termination of the multi-start, the best obtained solution S' replaces S in the population. In the second step, the current population is recombined to potentially construct new high quality solutions. In each recombination step, every solution in pop is combined with a randomly chosen second solution while ensuring that each pair of solutions is considered at most once. Given, such a pair of solutions S_1 and S_2 to create new solution the STP is solved (either heuristically or exactly) on the union of the subgraphs defined by the solutions. To prevent repetitive calculations of exact solutions of subinstances that have been treated before, MPCH also employs a solution archive storing optimal solutions of previously solved subinstances.

Variable neighborhood descent. Within variable neighborhood descent, we consider the following four neighborhood structures (in the same order as presented) from the literature using fast neighborhood evaluation recently proposed by Uchoa and Werneck [10]:

- The *Steiner node insertion* neighborhood structure contains all solutions which can be constructed from an initial solution S through the insertion of a single Steiner nodes $v \notin V_S$ (and edges $\{v, u\}$, $u \in V_S$) and the application of MST-Prune to the induced subgraph.
- The *key-path exchange* neighborhood structure contains all solutions which can be constructed from a given solution S through exchanging a key path P_1 from S by a new key path connecting the two components of S obtained after removing P_1. Thereby, a path is called a key path if its endpoints are either terminal nodes or Steiner nodes with degree at least three (in S) and all inner nodes of the path are Steiner nodes of degree two.
- The *key-node elimination* neighborhood structure is defined by all solutions that can be created from a given solution S by removing a single key node (i.e., a Steiner node of degree at least three) as well as its incident key paths, and reconnecting the resulting subtrees through a set of shortest paths.

4 Computational Results

In this section, our computational results are presented and analyzed. All algorithms are implemented in C++ and compiled using GCC 4.8.1 with the full compiler optimization flag (-O4). The B&C approach for model (EDCF) builds upon IBM CPLEX 12.5, METIS [14] is used for computing a k-way graph partition, and bossa [18] for preprocessing of STP instances. The test runs of all experiments for which we report runtimes, have been computed on a single core of an Intel Xeon E5540 2.53 GHz with 24 GB RAM.

Test instances have been selected by multiple criteria. We focused on large, sparse instances, since these are the primary focus of our algorithm. Sets ES, VLSI, and TSPFST have been chosen from the SteinLib [19]. The set VLSI contains 10 instances with a low ratio of terminals, while ES and TSPFST contain 16 and 10 instances, respectively, with a relatively high number of terminals, see [11] for more details. In addition, we use new large-scale real-world instances (10 particularly large instances from set I and 22 instances from set GEO) from telecom applications [20]. These are on average larger than the instances contained in the SteinLib, with up to 70 000 nodes and 110 000 edges after preprocessing.

4.1 Evaluation of the Partitioning Algorithms

We first evaluate the proposed *partitioning algorithms*. For comparison purposes, each algorithm is applied together with the *heuristic repair* algorithm, since this configuration is expected to require the lowest runtime. This choice does not have any impact on the performance of partitioning, since the application of these techniques is independent from each other.

Table 1 details the solution quality obtained from the different partitioning schemes. Each column contains results for a different partitioning scheme and the given parameters k and d. The following abbreviations are used: Eb and Vb denote the edge-based and Voronoi-based partitioning schemes, c_u and c_{orig} are the uniform and original weighting strategies for Eb, while S_G denotes the use of a guiding solution. This guiding solution is constructed by applying SPH and MST-Prune in the auxiliary graph generated by dual ascent. Note that instance set GEO is not considered in these initial experiment, since the structure of these instances is quite similar to the one of instances from set I.

We note that the results confirm that the parameters k and d affect PCH as intended. In most cases, a clear progression is visible concerning runtime (cf. [11]) and solution quality when the values of $|T|/k$ and d are increased. The creation of larger, imbalanced subinstances yields an improved solution quality, but increases the runtime, since the created subinstances take longer to be solved. Furthermore, we observe that Vb outperforms Eb with respect to the obtained solution quality in almost all cases. In addition, the solution quality obtained when using Eb heavily depends on the chosen parameter values which is not true for Vb. Our results also clearly indicate that the edge-cut of a graph is not a good measure to encourage the construction of good quality solutions. For Vb we observe that increasing d generally leads to better results. The exception is the

Table 1. Comparing partitioning schemes in PCH w.r.t. the average gaps between the computed solutions and the known optimum for each method [%]. Smallest average gaps per considered setting are marked bold.

	$|T|/k$	d	Edge-based partitioning (METIS)				New Voronoi-based partitioning	
			c_u	c_{orig} c_u, S_G		c_{orig}, S_G	$-$	S_G
ES	10	1	8.88	9.13	8.57	8.13	1.54	**0.98**
		2	1.99	1.43	1.37	1.05	1.03	**0.69**
		3	1.82	1.24	1.29	0.94	0.97	**0.63**
	100	1	3.41	3.90	1.35	1.53	0.37	**0.18**
		2	0.67	0.64	0.58	0.42	0.22	**0.14**
		3	1.63	0.95	0.80	0.75	0.20	**0.13**
VLSI	10	1	55.39	44.57	45.02	42.67	3.30	**1.25**
		2	8.65	7.95	5.22	4.62	1.29	**0.97**
		3	11.82	12.92	6.61	6.74	1.00	**0.81**
	100	1	25.50	25.15	19.99	20.09	1.86	**0.65**
		2	3.30	1.61	1.34	1.14	0.63	**0.58**
		3	3.27	1.46	1.23	1.11	0.71	**0.68**
TSPFST	10	1	7.42	7.33	8.50	7.50	1.69	**1.12**
		2	1.82	1.34	1.37	1.14	1.07	**0.82**
		3	1.80	1.30	1.33	1.12	0.98	**0.81**
	100	1	4.55	4.24	4.03	4.02	0.43	**0.28**
		2	0.59	0.53	0.40	0.73	0.35	**0.19**
		3	0.57	0.34	0.41	0.35	0.23	**0.14**
I	10	1	0.903	0.863	0.986	0.826	0.094	**0.039**
		2	0.241	0.198	0.144	0.125	0.056	**0.028**
		3	0.234	0.201	0.149	0.112	0.046	**0.028**
	100	1	0.340	0.296	0.320	0.281	0.094	**0.011**
		2	0.090	0.109	0.058	0.083	0.038	**0.006**
		3	0.097	0.076	0.065	0.111	0.016	**0.006**

VLSI instance set, for which $d = 2$ achieves the best results. We conclude that in instances which contain only a very small percentage of terminals, making subsets in a partition too big may lead to less favorable results than creating smaller subsets and connecting them heuristically. We also note that the average runtime of Vb is typically lower than the one of Eb and even for those cases where Eb was faster the difference was typically less than a second. We note that the running times of Vb are significantly larger than those of Eb only when using large values of d on instance set I. This is, however, compensated by better solution quality. We further note that using a guiding solution speeds up Vb for all considered parameter values. Overall, we conclude, that Vb performs much better both with respect to solution quality and runtime, and is also a much more robust strategy with respect to the choice of parameters. Thus, only Vb with a guiding solution is used when referring to PCH in the remaining experiments.

4.2 Comparison to Other Methods

Tables 2 and 3 compare average gaps to the best known objective values (which are proven optimal values except for several of the GEO instances) and average runtimes grouped by instance set, see also Figure 3 for a graphical representation. In the listed results *SPH+VND* refers to applying the shortest path heuristic (SPH) followed by VND, and *DA+VND* to a variant of SPH+VND applied to the auxiliary graph which remains after execution of the dual ascent algorithm. The latter may not only provide better solutions but more importantly also provides a lower bound for the estimation of the solution's quality. *PCH+VND* refers to a using PCH as initial solution for VND using the best parameters from preliminary experiments (i.e., Voronoi-based partitioning with $k = |T|/100$, $d = 3$, guiding solution, and heuristic repair), see also [11]. Thereby, the guiding solution is produced as in DA+VND. *MPCH* is the memetic algorithm with parameters $g = 3$, *popmax* $= 10$, $n = 2$, which have been determined in preliminary experiments, see also [11], and using the exact solution recombination. PCH is internally applied as in PCH+VND.

In all variants, the time limit for each exact solution of a subinstance is set to $t = 100s$. Finally, *HGPPR* refers to the hybrid GRASP with perturbations and path relinking proposed by Ribeiro et al. [9] which has been rerun on our environment. The publicly available implementation [18] has been configured as follows: The number of iterations for the GRASP is fixed to 128. For the path relinking phase, no restriction is enforced, and the algorithm only terminates if no improved solution can be found based on the current population. Adaptive path relinking is used, which means that the program tests the runtime of two different path relinking algorithms for a few iterations, and chooses the faster one. In the following, we present both solution quality and runtime for only the GRASP phase and also the full procedure. The results after the GRASP phase are denoted by HGP, while the result of the full procedure are denoted by HGPPR. We note that although HGPPR has been applied to all instance sets in our experiments, we have been unable to record any meaningful results for the instance set GEO, since the used implementation failed to process this type of instance correctly. In addition to the heuristic approaches, we also show the results for the B&C with a time limit of 24 hours. Note that we did not compare to the state-of-the-art exact approach of Polzin [4] and Daneshmand [6] since their implementation is not publicly available.

We observe that the combination of SPH and VND produces good quality solutions fast, even for larger instances. The dual ascent implementation typically yields slightly better solutions but needs significantly more time for large-scale instances. We see that a single iteration of PCH with local search yields excellent results in a small amount of time and that for the large instances with many terminals, a single iteration of PCH yields a better bound than HGPPR. The required runtime is also extremely low, in particular compared to the long runtimes of HGPPR. We conclude that PCH outperforms the other algorithms if there are many terminal nodes.

Table 2. Comparison of different methods w.r.t. the average gaps between the obtained and the best known solution values [%]. Smallest average gaps among the VND-based methods and among the other methods marked bold.

	SPH+VND	DA+VND	PCH+VND	MPCH	HGP	HGPPR	B&C
ES	0.82	0.57	**0.12**	0.04	0.32	0.09	**0.005**
TSPFST	0.88	0.70	**0.14**	**0.03**	0.31	0.09	0.037
VLSI	1.41	0.99	**0.33**	0.17	0.30	**0.10**	0.186
I	0.0264	0.0167	**0.0019**	0.0008	0.0100	0.0044	**0.0000**
GEO	0.9815	0.8595	**0.2654**	0.0681	-	-	**0.0478**

Table 3. Comparison of different methods w.r.t. the average runtimes [s]. Smallest average runtimes among the VND-based methods and among the other methods are marked bold.

	SPH+VND	DA+VND	PCH+VND	MPCH	HGP	HGPPR	B&C
ES	**0.08**	0.21	4.47	69.55	148.51	365.96	**56.11**
TSPFST	**0.13**	0.26	19.16	261.02	167.29	380.07	201.54
VLSI	**0.11**	0.39	37.77	117.63	**47.33**	76.76	458.13
I	**4.14**	31.31	156.95	**1621.09**	6262.05	63134.15	21906.91
GEO	**2.02**	10.78	200.28	**1515.36**	-	-	20229.57

Concerning MPCH, the number of iterations and exact recombination leads to a significant runtime increase compared to PCH. Despite the fact that MPCH is costly with respect to runtime, it clearly pays off regarding solution quality. Note that an improvement is achieved in all cases compared to PCH+VND. Moreover, MPCH also outperforms HGPPR in the majority of cases.

We note that the bossa framework containing HGPPR does not employ the improved local search strategies. They employ Steiner node insertion/elimination and key-path exchange but not their improved implementations [10] which yield a quite huge runtime for large-scale instances with many nodes. This clearly highlights the importance of the improved implementations of the neighborhood exploration. We assume that the runtime of HGPPR can be improved greatly through the application of these implementations. We also note that though B&C is an exact method, it is quite competitive to the other methods. The average runtimes for the sets ES, TSPFST and VLSI are not that far away from the ones of the more sophisticated heuristics considered. The worst performance is achieved for the VLSI instances which are known to be very hard for ILP based approaches due to their regular cost structure. For ES, the average runtime and gap are even better than for other approaches. The average runtime for the I instance set is large, but not as large as the one of HGPPR. Only MPCH seems

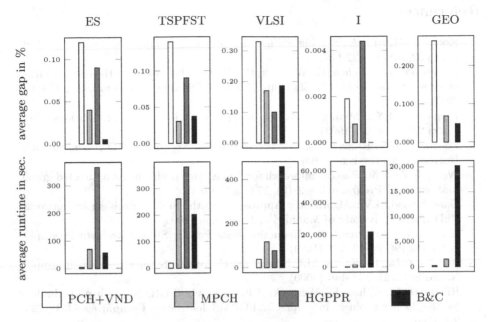

Fig. 3. A graphical representation of the performance comparison from Table 2 and Table 3 comparing PCH+VND, MPCH, HGPPR and B&C.

to achieve an average gap close to B&C. We therefore conclude, that B&C is a very powerful approach by itself, and that even if the time available for exact solution is limited, acceptable results can be achieved.

5   Conclusions

In this article, a new partition-based construction heuristic (PCH) for effectively solving large scale STP instances has been proposed. PCH combines a novel approach to STP graph partitioning with state-of-the-art exact and heuristic methods. The approach has also been incorporated into a partition-based memetic algorithm (MPCH). Computational experiments have been performed on standard benchmark instances from the literature and new real-world instances from telecommunications. The obtained results show that PCH is able to produce high quality solutions (with gaps close to zero) and is competitive to other state-of-the-art methods for the tested instances. In addition, the algorithm's runtime is orders of magnitude faster than the other methods for large-scale instances. The introduced concept of Voronoi-based partitioning clearly outperforms other tested variants and performs excellent in particular on sparse graphs and when the relative number of terminals is relatively high or when a natural clustering of terminals exists in an instance. Further improvements with respect to solution quality have been achieved in MPCH which come, however, at the price of a significantly higher runtime.

References

1. Koch, T., Martin, A.: Solving Steiner tree problems in graphs to optimality. Networks 32(3), 207–232 (1998)
2. de Aragão, M.P., Uchoa, E., Werneck, R.F.F.: Dual heuristics on the exact solution of large Steiner problems. Electronic Notes in Discrete Mathematics 7, 150–153 (2001)
3. Goemans, M.X., Myung, Y.S.: A catalog of Steiner tree formulations. Networks 23(1), 19–28 (1993)
4. Polzin, T.: Algorithms for the Steiner Problem in Networks. PhD thesis, Saarland University, Saarbrcken (2003)
5. Wong, R.T.: A dual ascent approach for Steiner tree problems on a directed graph. Mathematical Programming 28(3), 271–287 (1984)
6. Daneshmand, S.V.: Algorithmic Approaches to the Steiner Problem in Networks. PhD thesis, University of Mannheim, Mannheim (2003)
7. Duin, C.W., Volgenant, A.: Reduction tests for the Steiner problem in graphs. Networks 19(5), 549–567 (1989)
8. Ribeiro, C.C., De Souza, M.C.: Tabu search for the Steiner problem in graphs. Networks 36(2), 138–146 (2000)
9. Ribeiro, C.C., Uchoa, E., Werneck, R.F.F.: A hybrid GRASP with perturbations for the Steiner problem in graphs. INFORMS Journal on Computing 14(3), 228–246 (2002)
10. Uchoa, E., Werneck, R.F.F.: Fast local search for Steiner trees in graphs. In: Blelloch, G.E., Halperin, D. (eds.) ALENEX, pp. 1–10. SIAM (2010)
11. Luipersbeck, M.: A new partition-based heuristic for the Steiner tree problem in large graphs. Master's thesis, Faculty of Informatics, Vienna University of Technology (December 2013)
12. Guschinskaya, O., Dolgui, A., Guschinsky, N., Levin, G.: A heuristic multi-start decomposition approach for optimal design of serial machining lines. European Journal of Operational Research 189(3), 902–913 (2008)
13. Taillard, É.D., Voß, S.: POPMUSIC–Partial optimization metaheuristic under special intensification conditions. In: Essays and Surveys in Metaheuristics, pp. 613–629. Springer (2002)
14. Karypis, G., Kumar, V.: A fast and high quality multilevel scheme for partitioning irregular graphs. SIAM Journal on Scientific Computing 20(1), 359–392 (1998)
15. Cherkassky, B.V., Goldberg, A.V.: On implementing the push-relabel method for the maximum flow problem. Algorithmica 19(4), 390–410 (1997)
16. Takahashi, H., Matsuyama, A.: An approximate solution for the Steiner problem in graphs. Math. Japonica 24(6), 573–577 (1980)
17. Poggi de Aragão, M., Werneck, R.F.F.: On the implementation of MST-based heuristics for the Steiner problem in graphs. In: Mount, D.M., Stein, C. (eds.) ALENEX 2002. LNCS, vol. 2409, pp. 1–15. Springer, Heidelberg (2002)
18. Werneck, R.F.F.: Bossa (2003), http://www.cs.princeton.edu/~rwerneck/bossa/ (visited on January 20, 2014)
19. Koch, T., Martin, A., Voß, S.: SteinLib: An updated library on Steiner tree problems in graphs. Technical Report 00-37, ZIB, Takustr.7, 14195, Berlin (2000)
20. Leitner, M., Ljubić, I., Luipersbeck, M., Prossegger, M., Resch, M.: New real-world instances for the Steiner tree problem in graphs. Technical report, ISOR, University of Vienna (2014)

A Path-Generation Matheuristic for Large Scale Evacuation Planning

Victor Pillac[1], Pascal Van Hentenryck[1,2], and Caroline Even[1]

[1] NICTA Optimisation Research Group*,
Melbourne, Australia
{victor.pillac,pvh,caroline.even}@nicta.com.au
http://org.nicta.com.au
[2] Australian National University, Canberra, Australia

Abstract. In this study we present a general matheuristic that decomposes the problem being solved in a master and a subproblem. In contrast with the column generation technique, the proposed approach does not rely on the explicit pricing of new columns but instead exploits features of the incumbent solution to generate one or more columns in the master problem. We apply this approach to large scale evacuation planning, leading to the first scalable algorithm that complies with emergency services practice.

1 Introduction

Natural and man-made disasters, such as hurricanes, floods, bushfires, or industrial accidents, often affect large populated areas, threatening the lives and welfare of entire populations. In such events, a common contingency is to evacuate the persons at risk to different shelters and safe areas.

Existing work in evacuation planning typically relies on free-flow models in which evacuees are dynamically routed in the network. In contrast, this paper presents an evacuation algorithm that follows recommended evacuation methodologies, which divide the evacuated area in evacuation zones, each being instructed to leave at a specific time and following a pre-defined route [24]. More specifically, it generates evacuation routes for each evacuation zone and uses a lexicographic objective function that first maximizes the number of evacuees reaching safety and then minimizes the total evacuation time. The algorithm can be used for strategic and tactical planning.

From a technical standpoint, the algorithm can be broadly characterized as a Conflict-Based Path-Generation Heuristic (CPG for short), which shares some characteristics with column generation approaches. As in column generation, it decomposes the problem by considering separately the generation of evacuation paths (subproblem) and their selection (master problem). However, a challenge of

---

* NICTA is funded by the Australian Government through the Department of Communications and the Australian Research Council through the ICT Centre of Excellence Program.

M.J. Blesa, C. Blum, and S. Voß (Eds.): HM 2014, LNCS 8457, pp. 71–84, 2014.

our application is the spatio-temporal nature of the problem: one evacuation path corresponds to multiple paths in the spatio-temporal graph modeling the actual scheduling of the evacuation. As a consequence, the master problem contains interdependent binary selection variables and continuous flow variables for each evacuation path, which means that pricing a single column is not relevant, and we therefore focus on generating evacuation paths. We propose two approaches for the sub-problem. The first explicitly attempts to find a path that will improve the value of the master problem objective function, while the second aims at finding a path of least cost under constraints, where the edge costs are derived from the conflicts and congestion in the incumbent solution.

We evaluate the CPG algorithm on real-scale, massive flood scenarios in the Hawkesbury-Nepean river (West Sydney, Australia) which require evacuating in the order of 70,000 persons. Experimental results indicate that the CPG algorithm generates high-quality solutions in limited time. On small instances, where optimal solutions can be found, the CPG algorithm finds optimal or near-optimal solutions. On real-scale instances, the results show that the CPG algorithm is capable of evacuating the entire Hawkesbury-Nepean region in under 10h even if the population grows by 20%.

The remainder of this paper is organized as follows: Section 2 formulates the Evacuation Planning Problem (EPP), Section 3 reviews related work, Section 4 presents the solution approaches, Section 5 compares the performance of the proposed approaches on a set of realistic instances, and, finally, Section 6 concludes this paper.

2   Problem Formulation

Figure 1 illustrates an instance of the Evacuation Planning Problem (EPP). Fig. 1(a) presents an evacuation scenario with one evacuated node (0) and two safe nodes (A and B). In this example, the evacuated node 0 has to be evacuated by 13:00, considering that certain links become unavailable at different times (for instance, $(2, 3)$ is cut at 9:00). This evacuation scenario can be represented as a graph $\mathcal{G} = (\mathcal{N} = \mathcal{E} \cup \mathcal{T} \cup \mathcal{S}, \mathcal{A})$ where $\mathcal{E}$, $\mathcal{T}$, and $\mathcal{S}$ are the set of evacuated, transit, and safe nodes respectively, and $\mathcal{A}$ is the set of edges. Each evacuated node i is characterized by a number of evacuees d_i and an evacuation deadline $\bar{f}_i$ (e.g., 20 and 13:00 for node 0 respectively), while each edge e is associated with a triple $(s_e, u_e, \bar{f}_e)$, where s_e is the travel time, u_e is the capacity, and $\bar{f}_e$ is the time at which the edge becomes unavailable.

A common way to deal with the space-time aspects of evacuation problems is to discretize the planning horizon into time steps of identical length, and to work on a *time-expanded graph*, as illustrated in Fig. 2. This graph $\mathcal{G}^d = (\mathcal{N}^d = \mathcal{E}^d \cup \mathcal{T}^d \cup \mathcal{S}^d, \mathcal{A}^d)$ is constructed by duplicating each node from $\mathcal{N}$ for each time step. For each edge $(i, j) \in \mathcal{A}$ and for each time step t in which edge (i, j) is available, an edge $(i_t, j_{t+s_{(i,j)}})$ is created modeling the transfer of evacuees from node i at time t to node j at time $t + s_{(i,j)}$. In addition, edges with infinite capacity are added to model evacuees waiting at evacuated and safe nodes. Finally, all evacuated nodes (resp. safe nodes) are connected to a virtual

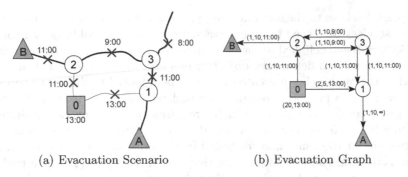

(a) Evacuation Scenario   (b) Evacuation Graph

Fig. 1. Modeling of an Evacuation Planning Problem

super-source v_s (super-sink v_t), modeling the inflow (outflow) of evacuees. Note that some nodes may not be connected to either the super-source or super-sink (in light gray in this example), and can therefore be removed to reduce the graph size. The problem is then to find a flow from v_s to v_t that models the movements of evacuees in space and time.

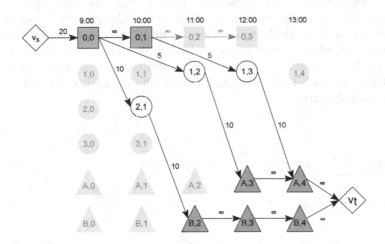

Fig. 2. Time-Expanded Graph for the Evacuation Scenario With 1-hour Time Steps

In this study, we will make the following assumptions:

1. A decision-maker instructs each evacuee when to leave, which safe node to go to, and which path to follow in the evacuation graph;
2. A single threat scenario is known at decision time;
3. The decision-maker objective is to ensure that all evacuees reach a safe node as early as possible;
4. Each evacuated node should be assigned to a single evacuation path;
5. Edge capacities do not depend on the flow, and no congestion occurs at intersections.

Assumption 1 relates to the fact that our approach is targeted toward emergency or safety services that need to design evacuation plans for buildings or regional areas. Assumption 2 is linked to the deterministic nature of our models and algorithms. Assumption 3 defines the objective function we will use, but it must be noted that the approaches presented here can be adapted to different contexts. Requirement 4 is a practical consideration and reflects the practice in the field of emergency services operations. Finally, requirement 5 is a necessary simplification to solve the models efficiently, and it is compensated by the fact that edge capacities are set to ensure non-congested flow conditions. In that context, the problem is to design an evacuation plan that assigns a single evacuation path to each evacuated node, and to schedule the evacuation over the planning horizon, with the objective of first maximizing the number of evacuees reaching a safe node, and then minimizing the time at which the last evacuee reaches safety.

Model (1-9) presents a Mixed Integer Program modeling the Evacuation Planning Problem (EPP-MIP). Let $x_{e_0}^k$ be a binary variable equal to 1 if and only if edge $e_0 \in \mathcal{A}$ belongs to the evacuation path for evacuated node k, and φ_e^k a continuous variable equal to the flow of evacuees from evacuated node k on edge $e \in \mathcal{A}^d$. Constraints (2) ensure that exactly one path is used to route the flow coming from a same evacuated node in the evacuation graph, while constraints (3) ensure the continuity of the path. Constraints (4) ensure the flow conservation through the time-expanded graph. Constraints (5) enforce the capacity of each edge in the time-expanded graph. Constraints (6) ensure that there is no flow of evacuees coming from an evacuated node k if edge e is not part of the evacuation path for k, and Constraints (7) ensure that all evacuees leave the virtual source.

$$\min \quad \sum_{e \in \mathcal{A}^d} c_e \varphi_e \tag{1}$$

$$\text{s.t.} \quad \sum_{e_0 \in \delta_0^+(k)} x_{e_0}^k = 1 \qquad \forall k \in \mathcal{E} \tag{2}$$

$$\sum_{e_0 \in \delta_0^-(i)} x_{e_0}^k - \sum_{e_0 \in \delta_0^+(i)} x_{e_0}^k = 0 \qquad \forall k \in \mathcal{E}, i \in \mathcal{T} \tag{3}$$

$$\sum_{e \in \delta^-(i)} \varphi_e^k - \sum_{e \in \delta^+(i)} \varphi_e^k = 0 \qquad \forall i \in \mathcal{N}^d \setminus \{v_s, v_t\}, k \in \mathcal{E} \tag{4}$$

$$\sum_{k \in \mathcal{E}} \varphi_e^k \le u_e \qquad \forall e \in \mathcal{A}^d \tag{5}$$

$$\varphi_e^k \le u_e * x_{e_0}^k \qquad \forall e \in \mathcal{A}^d, k \in \mathcal{E} \tag{6}$$

$$\varphi_{(v_s, k)} = d_k \qquad \forall k \in \mathcal{E} \tag{7}$$

$$\varphi_e^k \ge 0 \qquad \forall e \in \mathcal{A}^d, k \in \mathcal{E} \tag{8}$$

$$x_e^k \in \{0, 1\} \qquad \forall e \in \mathcal{A}^d, k \in \mathcal{E} \tag{9}$$

The objective (1) is to maximize the number of evacuees reaching a safe node and minimize the weighted evacuation time. For computational efficiency, we associate a penalty with edges arriving to a safe node proportional to the time slice in which they belong. Let $t(i)$ be the time slice of time-node i, and c_{ne} a high penalty for non-evacuated evacuees. The cost $c_{(i,j)}$ of edge $(i,j) \in \mathcal{A}^d$ is defined as:

$$c_{(i,j)} = \begin{cases} c_{ne} & \text{if } i \in \mathcal{E}^d, j = v_t \\ \frac{t(i)}{H} & \text{if } i \in \mathcal{T}^d, j \in \mathcal{S}^d \\ 0 & \text{otherwise} \end{cases} \qquad (10)$$

3 Related Work

According to Hamacher and Tjandra [10], evacuation planning can be tackled using either *microscopic* or *macroscopic* approaches. Microscopic approaches focus on modeling and simulating the evacuees individual behaviors, movements, and interactions. Macroscopic approaches, such as the one presented in this study, aggregate evacuees and model their movements as a flow in the evacuation graph.

To the best of our knowledge, only a handful of studies attempt to design evacuation plans as we defined them [22]. Huibregtse et al. [14] propose a two stage algorithm that first generates a set of evacuation routes and feasible departure time, and then assigns a route and time to each evacuated area using an ant colony optimization algorithm. A key feature of the approach is the use of traffic simulation to evaluate the quality of solutions. In later work, the authors studied the robustness of the produced solution [13], and strategies to improve the compliance of evacuees [12].

A significant number of contributions attempt to solve flow problems directly derived from the time-expanded graph. For instance, Lu et al. [18, 19] propose three heuristics to design an evacuation plan with multiple evacuation routes per evacuated node, minimizing the time of the last evacuation. The authors show that in the best case the proposed heuristic is able to solve randomly generated instances of up to 50,000 nodes and 150,000 edges in under 6 minutes. Liu et al. [17] propose a Heuristic Algorithm for Staged Traffic Evacuation (HASTE), a similar algorithm that generates augmenting chains in the time-expanded graph. The main difference between HASTE and the previous algorithms is that it relies on a Cell Transmission Model (CTM)[8] to model more accurately the flow of evacuees.

Acknowledging that all evacuated nodes may not be under the same level of threat, Lim et al. [15] consider a short-notice regional evacuation maximizing the number of evacuees reaching safety weighted by the severity of the threat. The authors propose two solution approaches to solve the problem, and present computational experiments on instances derived from the Houston-Galveston region (USA) with up to 66 nodes, 187 edges, and an horizon of 192 time steps.

Other authors have focused on modeling more accurately the transportation network. For example, Bretschneider and Kimms [5, 6] present a free-flow mathematical model that describes in detail the street network and, in particular, the lane configuration at intersections of the network. They present computational experiments on generated instances with a grid topology of up to 240 nodes, 330 edges, and considering 150 times steps. Bish and Sherali [4] present a model based on a CTM that assigns a single evacuation path to each evacuated node. Computational results include instances with up to 13 evacuated nodes, 2 safe nodes, and 72 edges.

Finally, dynamic aspects of evacuation have also been considered. For instance, Lin et al. [16] present a time expanded graph in which they allow for time-dependent attributes such as varying capacity or demand. The authors apply their findings on a case study considering the evacuation of a 11-floor building with approximately 60 nodes, 100 edges, and 60 time steps.

Microscopic approaches include the work by Richter et al. [23] who challenge two assumptions generally made: The existence of a central planning entity with global knowledge, and the ability of this entity to communicate order to evacuees. They propose a decentralized decision making approach supported by smartphones and mobile applications. We note however that our target applications, such as evacuations for floods and hurricanes, use central decision making and have the time and ability to communicate their decisions.

In contrast with the cited studies, the approach proposed in this work is the first to produce evacuation plans that are actionable from an emergency service perspective. It generates a plan that assigns one evacuation route to each evacuated area, and optimizes both the evacuation routes and schedule globally.

Column generation is an optimization technique which consists in considering only a subset of columns in a master problem and then iteratively generating columns of negative reduced cost (assuming minimization) by solving a pricing subproblem. It has been widely used to solve large-scale MIP problems, and we refer the interested reader to the book by Desaulniers et al. [9] and the study by Luebbecke and Desrosiers [20] for a recent review of techniques and applications of column generation. In particular, it has been used to solve multi-commodity network flow problems (MCNF) [1], integer MCNF [3], origin-destination MCNF [2], and MCNF with side constraints on paths [11].

However, a distinctive feature of evacuation planning is the dependency between paths in the time-expanded network. More precisely, a commodity (i.e., evacuees from a specific evacuated node) can only follow paths that correspond to the same physical path (sequence of edges in the evacuation graph). Therefore classical MCNF approaches cannot be applied directly, as one path in the evacuation model introduces multiple variables in the master problem. In addition, it is worth noting that heuristic column generation have mainly focused on solving the pricing subproblem heuristically. In contrast, our approach does not consider the pricing problem explicitly, but heuristically generates new paths. Similar ideas were also used by Coffrin et al. [7] and, to a lesser extent, in Massen et al. [21].

4 Proposed Approaches

Computational experiments show that the EPP-MIP model (1-9) becomes intractable for instances with more than eight nodes and an horizon of 10h divided in 5 minutes steps. Therefore, we propose a conflict-based heuristic path generation approach (CPG) that separates the generation of evacuation paths from the scheduling of the evacuation.

Figure 3 gives an overview of the CPG approach. First, the algorithm generates an initial set of paths Ω' (1) and solves a master problem to find an evacuation schedule optimizing the objective function (2). Then it identifies *critical* evacuated nodes $\mathcal{E}'$ (3), which are not fully evacuated, or evacuated late, and considers nodes that are potentially in conflict (5) with the objective of generating new paths (6). Finally, it solves the scheduling problem including the newly generated paths (2). The steps are repeated for a fixed number of iterations.

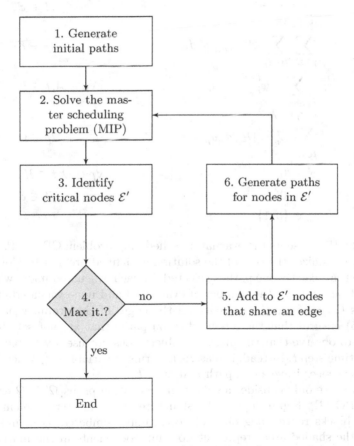

Fig. 3. Overview of the conflict-based heuristic path generation approach

4.1 Master Problem

Let Ω be the set of all feasible paths between evacuated nodes and safe nodes and Ω_k be the subset of evacuation paths for evacuated node k. We define a binary variable x_p which takes the value of 1 if and only if path $p \in \Omega$ is selected, a continuous variable φ_p^t representing the number of evacuees to start evacuating on path p at time t, and a continuous variable φ_k accounting for the number of non-evacuated evacuees in node k. In addition, we denote by $\omega(e)$ the subset of paths that contain edge e and by τ_p^e the number of time steps required to reach edge e when following path p. Finally, we denote by $\mathcal{H}_p \subseteq \mathcal{H}$ the subset of time steps in which path p is usable, and u_p the capacity of path p.

$$\min \quad \sum_{k \in \mathcal{E}} \varphi_k c_{ne} + \sum_{p \in \Omega} \sum_{t \in \mathcal{H}_p} \varphi_p^t c_p^t \tag{11}$$

$$\text{s.t.} \quad \sum_{p \in \Omega_k} x_p = 1 \qquad\qquad \forall k \in \mathcal{E} \tag{12}$$

$$\sum_{p \in \Omega_k} \sum_{t \in \mathcal{H}_p} \varphi_p^t + \varphi_k = d_k \qquad\qquad \forall k \in \mathcal{E} \tag{13}$$

$$\sum_{\substack{p \in \omega(e) \\ t - \tau_p^e \in \mathcal{H}_p}} \varphi_p^{t - \tau_p^e} \leq u_e \qquad\qquad \forall e \in \mathcal{A}, t \in \mathcal{H} \tag{14}$$

$$\sum_{t \in \mathcal{H}_p} \varphi_p^t \leq |\mathcal{H}_p| x_p u_p \qquad\qquad \forall p \in \Omega \tag{15}$$

$$\varphi_p^t \geq 0 \qquad\qquad \forall p \in \Omega, t \in \mathcal{H}_p \tag{16}$$

$$\varphi_k \geq 0 \qquad\qquad \forall k \in \mathcal{E} \tag{17}$$

$$x_p \in \{0, 1\} \qquad\qquad \forall p \in \Omega \tag{18}$$

Model (11-18) presents the evacuation scheduling problem CPG-MP. The objective (11) minimizes the cost of the solution as defined previously. Constraints (12) ensure that exactly one path is selected for each evacuated node, while constraints (13) account for the number of evacuated and non-evacuated evacuees. Constraints (14) enforce the capacity on the edges of the graph. Finally, constraints (15) ensures that there is no flow on paths that are not selected. It is interesting to observe that the master problem does not use a variable for each edge e and time step t. Instead, it reasons in terms of variables φ_p^t which indicate how many evacuees leave along path p at time t.

In practice, we only consider a subset of evacuation paths $\Omega' \subseteq \Omega$ each time we solve CPG-MP. Fig. 4 depicts the structure of the master problem matrix. Horizontal blocks represent groups of constraints numbered as in model (11-18), while the shaded areas represent non-null coefficients in the matrix. Note that each constraint in group (15) only involves variables associated with the corresponding path and must be dynamically added to the model whenever a new path is considered. Nonetheless, a solution of CPG-MP considering the

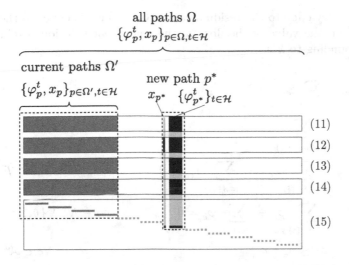

Fig. 4. The Structure of the Evacuation Scheduling Master Problem Matrix

subset of paths $\Omega'' \subset \Omega'$ is feasible when considering the set Ω'. Hence the solution from the previous iteration is used as starting solution for the current iteration.

4.2 Subproblem

Considering the spatio-temporal nature of this application, and the fact that a path corresponds to multiple columns and introduces a new constraint, we do not rely on traditional column generation techniques to generate new paths. Instead, we use problem-specific knowledge to generate new columns that will potentially improve the objective function of the master problem. First, we identify the subset $\mathcal{E}' \subseteq \mathcal{E}$ of critical evacuated nodes, i.e., nodes that are not fully evacuated in the current solution or evacuated late. Then, we include in $\mathcal{E}'$ all the evacuated nodes whose evacuation paths share at least one edge with a node from $\mathcal{E}'$. Finally, we generate new paths for the critical evacuated nodes $\mathcal{E}'$.

Improving Path Generation (IPG). The first path generation approach we propose borrows ideas from the Large Neighborhood Search algorithm [25]. Conceptually, we transform a solution of CPG-MP into a solution of EPP-MIP, fixing all flow variables in EPP-MIP to their value derived from CPG-MP, except for the subset of variables corresponding to the evacuation area $k^* \in \mathcal{E}'$ for which a new path is to be generated. In other words, we solve EPP-MIP for a single evacuated area (k^*) reducing the capacity on the edges of the time-expanded graph. Model (19-26) presents a MIP formulation of the resulting problem. The model is very similar to the original problem, at the difference that only one evacuation area is considered. In addition, Constraints (23) limit the flow on every edge to the residual capacity u'_e which is equal to the original capacity minus the value of the flow on the incumbent solution (excluding the flow corresponding to k^*).

$$\min \quad \sum_{e \in \mathcal{A}^d} c_e \varphi_e \tag{19}$$

$$\text{s.t.} \quad \sum_{e_0 \in \delta_0^+(k^*)} x_{e_0} = 1 \tag{20}$$

$$\sum_{e_0 \in \delta_0^-(i)} x_{e_0} - \sum_{e_0 \in \delta_0^+(i)} x_{e_0} = 0 \qquad \forall k \in \mathcal{E}, i \in \mathcal{T} \tag{21}$$

$$\sum_{e \in \delta^-(i)} \varphi_e - \sum_{e \in \delta^+(i)} \varphi_e = 0 \qquad \forall i \in \mathcal{N}^d \setminus \{v_s, v_t\} \tag{22}$$

$$\varphi_e \leq u'_e * x_{e_0} \qquad \forall e \in \mathcal{A}^d \tag{23}$$

$$\varphi_{(v_s, k^*)} = d_{k^*} \tag{24}$$

$$\varphi_e \geq 0 \qquad \forall e \in \mathcal{A}^d \tag{25}$$

$$x_e \in \{0, 1\} \qquad \forall e \in \mathcal{A}^d \tag{26}$$

The values of the $(x_{e_0})_{e_0 \in \mathcal{E}}$ variables in each solution of the IPG problem define a new path for the CPG-MP. If the IPG problem does not admit a solution, or if the value of the solution is greater than the contribution of k^* to the incumbent solution value, we select up to $n = 5$ evacuation areas which share some edges with k^* and remove their contribution to residual capacity. This is equivalent to assuming that the removed areas do not need to be evacuated, and it relaxes the IPG by increasing the capacity on the edges. This process is repeated for all evacuated nodes identified as critical.

Heuristic Path Generation (HPG). The second path generation approach attempts to generate diverse evacuation paths by solving a series of independent shortest path problems from each evacuated node to all safe nodes. The cost c_e of edge e is adjusted at each iteration using the following linear combination of the edge's travel time s_e, the number of occurrences of e in the current set of paths, and the utilization of e in the current solution:

$$c_e = \alpha_t \frac{s_e}{\max_{e \in \mathcal{E}} s_e} + \alpha_c \frac{\sum_{p \in \Omega'} \underset{e \in p}{1}}{|\Omega'|} + \alpha_u \frac{\sum_{p \in \Omega'} \sum_{t \in \mathcal{H}_p} \varphi_p^t}{u_e} \tag{27}$$

where α_t, α_c, and α_u are positive weights which sum is equal to 1.

5   Computational Experiments

We consider the evacuation of the Hawkesbury-Nepean (HN) floodplain, located North-West of Sydney, for which a 1-in-200 years flood will require the evacuation of 70,000 persons. The resulting evacuation graph contains 50 evacuated nodes, 6 safe nodes, 153 transit nodes, and 485 edges. We consider a horizon of 10 hours with a time step of 5 minutes (starting at 00h00). In addition, we generate the instances HN-Rx with a subset of $x \in [2, 50]$ evacuation nodes and a reduced graph, and HN-Ix which have the same evacuation graph but a number of evacuees scaled by a factor of $x \in [1.1, 3.0]$.

All approaches were implemented using Java 7 and GUROBI 5.5, and experiments were conducted on a cluster of 64bits machines with 2.8GHz AMD 6-Core Opteron 4184 and 16Gb of RAM. Results are an average over 10 runs given the randomized nature of parts of the algorithms and of GUROBI internal heuristics. We set a limit of 10 iterations for CPG, which generally converges quickly. The IPG subproblem is solved using the GUROBI solver, while HPG relies on the Dijsktra algorithm to evaluate the shortest paths.

Table 1 compares the percentage of evacuees reaching safety (Perc. Evac) and the time at which the last evacuees reaches safety (Evac. End) in the solutions produced by CPG and by solving the EPP-MIP with the GUROBI solver. The figures in bold indicate proved optimum solution, figures in italics represent incumbent solution at the 30 min time limit. The results indicate that for instances of up to 5 nodes, the three approaches find the optimal solution. However, for larger instances, EPP-MIP does not terminate in the time limit.

Table 1. Comparison of solution quality on reduced size instances

| | CPG | | | | EPP-MIP | |
| | IPG | | HPG | | | |
	Perc. Evac	Evac. End	Perc. Evac	Evac. End	Perc. Evac	Evac. End
HN-R02	100%	02h30	100%	02h30	**100%**	**02h30**
HN-R03	100%	01h55	100%	01h55	**100%**	**01h55**
HN-R05	100%	02h25	100%	02h25	**100%**	**02h25**
HN-R08	100%	02h50	100%	02h50	*100%*	*02h50*
HN-R10	100%	02h50	100%	02h50	*100%*	*04h25*
HN-R20	100%	02h25	100%	02h50	*78%*	*10h00*
HN-R30	100%	02h45	100%	03h05	*82%*	*10h00*
HN-R40	100%	09h15	100%	09h15	76%	10h00
HN-R50	100%	09h15	100%	09h15	-	-

Table 2 presents the percentage of evacuees reaching safety (Perc. Evac), the time at which the last evacuees reaches safety (Evac. End) and the number of paths generated (Num. Paths) for both the IPG and HPG path generation. For comparison, we include results obtained when solving the master problem with only the three shortest paths from each evacuated area to the closest safe nodes. The results show that for instances with up to 20% additional evacuees, the whole area can be evacuated in under 10h by both approaches. It is worth noting that on these instances IPG and HPG give the same solutions, but IPG generates more than 10 times less paths. When the population increases, HPG performs increasingly better than IPG, evacuating more people in the 10h limit. Both approaches dominate the 3 shortest paths baseline.

Table 2. Comparison of solution quality

	HPG			IPG			3 Shortest Paths		
Instance	Perc. Evac.	Evac. Time	Num. Paths	Perc. Evac.	Evac. Time	Num. Paths	Perc. Evac.	Evac. Time	Num. Paths
HN-I	100%	08h05	1057	100%	08h05	65	100%	09h55	150
HN-I1.1	100%	08h45	1058	100%	09h10	80	100%	09h55	150
HN-I1.2	100%	09h25	1117	100%	09h25	79	98%	10h00	150
HN-I1.4	99%	10h00	741	98%	10h00	69	95%	10h00	150
HN-I1.7	97%	10h00	1211	93%	10h00	80	84%	10h00	150
HN-I2.0	94%	10h00	1118	91%	10h00	85	75%	10h00	150
HN-I2.5	84%	10h00	1193	74%	10h00	87	65%	10h00	150
HN-I3.0	75%	10h00	995	61%	10h00	85	58%	10h00	150
Average	94%	09h31	1061	90%	09h35	79	84%	09h58	150

Table 3 provides further insights on the relative performance of both approaches. It presents the CPU time (in seconds) spent in the master problem (MP) and path generation subproblem (SP) for the different instances. It appears that in the IPG approach solving the master problem takes less than 10 second in total, while it requires 67 minutes in the HPG. This difference can be explained by the smaller number of paths generated by IPG, which greatly reduces the size of the MP. On the other hand, IPG spends 285 minutes on the generation of new paths, compared with 2 seconds for HPG. This is explained by the algorithm used to solve the subproblem: HPG relies on the Dijsktra algorithm which is polynomial in the size of the time expanded graph, while IPG uses a MIP solver.

Table 3. Comparison of average CPU times (in seconds) for the master problem (MP) and subproblem (SP)

	HPG		IPG	
Instance	MP	SP	MP	SP
HN	2,617	1	5	2,533
HN-I1.1	2,684	1	9	15,094
HN-I1.2	2,652	1	9	8,622
HN-I1.4	1,459	1	3	11,868
HN-I1.7	3,623	2	5	17,367
HN-I2.0	5,156	2	9	17,121
HN-I2.5	7,000	2	9	25,357
HN-I3.0	6,906	2	7	39,195
Average	4,012	2	7	17,145

6   Conclusions

In this paper, we presented a generic matheuristic which borrows ideas from column generation and large neighborhood search. This general framework decomposes a complex problem in a master and subproblem. Similarly to column generation approaches, the master problem provides information to the subproblem to generate new variables. The first major difference is that the subproblem does not seek to generate a single column, but a set of columns linked by a constraint. The second major difference is that the subproblem does not explicitly minimize the reduced cost of the new columns, which means that it does not require access to the dual variables of the master problem. This is particularly an advantage when the master problem is a MIP. Instead, it attempts to identify features of the current incumbent that can be improved to lead to a better solution.

We applied the proposed matheuristic to the planning of large scale evacuations and demonstrated that it was able to produce high quality evacuation plans in reasonable time. In that application, the master problem selects evacuation paths and schedule the flow of evacuees, while the subproblem generates new paths for nodes identified as critical. We compared two approaches to solve the subproblem. The first, namely IPG, transforms the incumbent solution of the master problem in a solution of the original problem, and relaxes the variables corresponding to the evacuation node for which a new evacuation is to be generated. The second, HPG, relies on a randomized heuristic that generates new paths solving a shortest path problem on the evacuation graph where edges are penalized depending on their usage in the incumbent solution.

Preliminary computational results indicate that HPG has an advantage in terms of solution quality and CPU time. However, it appears that IPG produces competitive results with significantly less paths, which translates into lower CPU times for the master problem.

Future work will focus on the development of ad-hoc algorithms to solve the IPG problem more efficiently to speed up the algorithm. In addition, we are investigating ways to control the set of paths included in the master problem to improve the solution quality without affecting the computational time.

References

[1] Alvelos, F., Valrio De Carvalho, J.: Solving multicommodity flow problems with branch-and-price. Technical Report (2000)

[2] Barnhart, C., Hane, C., Vance, P.: Integer multicommodity flow problems. Lecture Notes in Economics and Mathematical Systems, vol. 450, pp. 17–31 (1997)

[3] Barnhart, C., Hane, C.A., Vance, P.H.: Using branch-and-price-and-cut to solve origin-destination integer multicommodity flow problems. Operations Research 48(2), 318–326 (2000)

[4] Bish, D.R., Sherali, H.D.: Aggregate-level demand management in evacuation planning. European Journal of Operational Research 224(1), 79–92 (2013)

[5] Bretschneider, S., Kimms, A.: A basic mathematical model for evacuation problems in urban areas. Transportation Research Part A: Policy and Practice 45(6), 523–539 (2011)

[6] Bretschneider, S., Kimms, A.: Pattern-based evacuation planning for urban areas. European Journal of Operational Research 216(1), 57–69 (2012)

[7] Coffrin, C., Van Hentenryck, P., Bent, R.: Strategic stockpiling of power system supplies for disaster recovery. In: 2011 IEEE Power and Energy Society General Meeting, pp. 1–8. IEEE (2011)

[8] Daganzo, C.F.: The cell transmission model: A dynamic representation of highway traffic consistent with the hydrodynamic theory. Transportation Research Part B: Methodological 28(4), 269–287 (1994)

[9] Desaulniers, G., Desrosiers, J., Solomon, M.M. (eds.): Column Generation. Mathematics of Decision Making. Springer (2005)

[10] Hamacher, H.W., Tjandra, S.A.: Mathematical modelling of evacuation problems: A state of art. Tech. rep., Fraunhofer Institut für Techno und Wirtschaftsmathematik (2001)

[11] Holmberg, K., Yuan, D.: A multicommodity network-flow problem with side constraints on paths solved by column generation. INFORMS Journal on Computing 15(1), 42–57 (2003)

[12] Huibregtse, O., Hegyi, A., Hoogendoorn, S.: Blocking roads to increase the evacuation efficiency. Journal of Advanced Transportation 46(3), 282–289 (2012)

[13] Huibregtse, O.L., Bliemer, M.C., Hoogendoorn, S.P.: Analysis of near-optimal evacuation instructions. Procedia Engineering 3, 189–203 (2010); 1st Conference on Evacuation Modeling and Management

[14] Huibregtse, O.L., Hoogendoorn, S.P., Hegyi, A., Bliemer, M.C.J.: A method to optimize evacuation instructions. OR Spectrum 33(3), 595–627 (2011)

[15] Lim, G.J., Zangeneh, S., Baharnemati, M.R., Assavapokee, T.: A capacitated network flow optimization approach for short notice evacuation planning. European Journal of Operational Research 223(1), 234–245 (2012)

[16] Lin, P., Lo, S., Huang, H., Yuen, K.: On the use of multi-stage time-varying quickest time approach for optimization of evacuation planning. Fire Safety Journal 43(4), 282–290 (2008)

[17] Liu, H.X., He, X., Ban, X.: A cell-based many-to-one dynamic system optimal model and its heuristic solution method for emergency evacuation. In: Proc. 86th Annual Meeting Transportation Res. Board (2007)

[18] Lu, Q., George, B., Shekhar, S.: Capacity constrained routing algorithms for evacuation planning: A summary of results. In: Medeiros, C.B., Egenhofer, M., Bertino, E. (eds.) SSTD 2005. LNCS, vol. 3633, pp. 291–307. Springer, Heidelberg (2005)

[19] Lu, Q., Huang, Y., Shekhar, S.: Evacuation planning: A capacity constrained routing approach. In: Chen, H., Miranda, R., Zeng, D.D., Demchak, C.C., Schroeder, J., Madhusudan, T. (eds.) ISI 2003. LNCS, vol. 2665, pp. 111–125. Springer, Heidelberg (2003)

[20] Lübbecke, M., Desrosiers, J.: Selected topics in column generation. Operations Research 53(6), 1007–1023 (2005)

[21] Massen, F., Deville, Y., Van Hentenryck, P.: Pheromone-based heuristic column generation for vehicle routing problems with black box feasibility. In: Beldiceanu, N., Jussien, N., Pinson, É. (eds.) CPAIOR 2012. LNCS, vol. 7298, pp. 260–274. Springer, Heidelberg (2012)

[22] Pillac, V., Hentenryck, P.V., Even, C.: A conflict-based path-generation heuristic for evacuation planning. Tech. Rep. VRL-7393, NICTA, arXiv:1309.2693 (2013)

[23] Richter, K.F., Shi, M., Gan, H.S., Winter, S.: Decentralized evacuation management. Transportation Research Part C: Emerging Technologies 31, 1–17 (2013)

[24] SES-NSW: Hawkesbury nepean flood emergency sub plan. Tech. rep., State Emergency Service - New South Wales (2005)

[25] Shaw, P.: Using constraint programming and local search methods to solve vehicle routing problems. In: Maher, M.J., Puget, J.-F. (eds.) CP 1998. LNCS, vol. 1520, pp. 417–431. Springer, Heidelberg (1998)

A Variable Neighborhood Search Using Very Large Neighborhood Structures for the 3-Staged 2-Dimensional Cutting Stock Problem

Frederico Dusberger and Günther R. Raidl

Institute of Computer Graphics and Algorithms
Vienna Unviersity of Technology
Favoritenstr. 9/1861, 1040 Vienna, Austria
{dusberger,raidl}@ads.tuwien.ac.at

Abstract. In this work we consider the 3-staged 2-dimensional cutting stock problem, which appears in many real-world applications such as glass and wood cutting and various scheduling tasks. We suggest a variable neighborhood search (VNS) employing "ruin-and-recreate"-based very large neighborhood searches (VLNS). We further present a polynomial-sized integer linear programming model (ILP) for solving the subproblem of 2-staged 2-dimensional cutting with variable sheet sizes, which is exploited in an additional neighborhood search within the VNS. Both methods yield significantly better results on about half of the benchmark instances from literature than have been published before.

1 Introduction

Cutting and packing problems are among the most well-studied combinatorial optimization problems in literature. This is due to the versatility of these problems allowing many real-world applications to be modelled as such. In fact, both cutting and packing usually refer to one and the same problem, however it is common, according to the context, to use either one term or the other. Examples of applications include actual industrial glass, paper or steel cutting, container loading, VLSI design, or various scheduling tasks [1,2]. Consequently, there are many different variants of the basic cutting and packing problems that have been discussed and for which a multitude of approaches already exists. Wäscher et al. [3] present an extensive typology of these problems, as well as a literature review of the most important works for the different variants. Nevertheless, this research area remains interesting, as additional modifications and side constraints arise, especially promoted due to new requirements from industry. Therefore, finding a most economical solution is still a challenging goal which calls for new approaches that are especially tailored and capable of respecting these new side constraints.

In this paper we consider in particular the 2-dimensional cutting stock problem (2CS), which – being a variant of the classical bin packing problem (1BP, or 2BP respectively) – is NP-hard [4]. In the basic problem setting one is given

M.J. Blesa, C. Blum, and S. Voß (Eds.): HM 2014, LNCS 8457, pp. 85–99, 2014.

a set of rectangular elements which need to be cut from a minimal number of larger stock sheets. We also consider the common restriction that only *guillotine* cuts are allowed, i.e. cuts are always parallel to one of the sheet sides and reach from one border to the opposite one. A further common side constraint due to the restrictions of real-world cutting machines, is that the cutting has to be done using a certain number of *stages*. A k-staged cutting is a sequence of k stages of cuts, where each stage consists of a series of parallel guillotine cuts performed on the pieces obtained from the previous stage. The direction of the cuts in one stage is always orthogonal to the cuts in the previous stage. Here, we want to focus on 3-staged cutting, as this is a typical restriction, e.g. in the glass manufacturing industry [5,6]. Secondly, experimental studies have shown that frequently no substantial gains can be obtained from more stages [7]. In case of the strip packing problem without allowing rotation the asymptotic performance ratio of a 3-staged cutting approximating an optimal cutting has been shown to be 1.69103 [8], whereas the 2-staged case is unbounded. The step from 2-staged to 3-staged cutting unfortunately dramatically increases the practical difficulty of the problem due to the possibility of stacking elements arbitrarily within a strip. In existing approaches there are often restrictions to the way elements can be stacked [9,10], but often the solution quality can be significantly increased when dropping those restrictions.

In this work we present a variable neighborhood search (VNS) for the 3-staged 2CS which is composed by several very large neighborhood structures based on the "ruin-and-recreate" principle. Trying to further improve the solution quality we developed a new integer linear programming (ILP) model which also was applied in the reconstruction phase of a "ruin-and-recreate" neighborhood.

The next section provides a detailed problem description of the 2CS, followed by a literature review of the work related to our specific problem in section 3. In section 4 we present our VNS framework, with the basic large neighborhood searches. Section 5 describes the ILP model and how it is exploited in an additional large neighborhood search. Section 6 gives experimental results for the developed methods and an analysis thereof, and section 7 concludes this work.

2   Problem Definition

In the 2CS we are given a set of m rectangular *elements* $M = \{E_1, \ldots, E_m\}$ with dimensions $(h_1, w_1), \ldots, (h_m, w_m)$, also called *demand*, which can be grouped into $t \leq m$ *element types* having the same dimensions. Furthermore, we have a (potentially unlimited) stock of identical rectangular sheets of height $H > 0$ and width $W > 0$.

The objective is to find a *cutting pattern*, i.c. an arrangement of the elements in M on the stock sheets without overlap, s.t. the number of required sheets is minimal and the pattern can be cut in a 3-staged process. Elements are rotatable by $90°$, which is reflected by adding for each element $E_i \in M$ an element E_{m+i} with $(h_{m+i}, w_{m+i}) = (w_i, h_i)$. We assume that an instance is generally feasible, i.e. $0 < h_i \leq H$ and $0 < w_i \leq W$ for each $E_i \in M$. Stage-1 and stage-3 cuts are always horizontal while stage-2 cuts are vertical.

We refer to the rectangles resulting from stage-1 cuts as *strips*, and the ones resulting from stage-2 cuts as *stacks*. We say that a cutting pattern is in *normal form*, if

(i) Waste occurs only at the bottom of stacks, at the right end of a strip and at the bottom of the sheet.
(ii) The topmost element of each stack is the widest one.
(iii) The leftmost stack of each strip is the highest one.

Clearly, every cutting pattern can be transformed into a pattern in normal form of equivalent quality. It is therefore sufficient to only consider patterns in normal form in the optimization.

We use the refined objective function proposed by Puchinger et al. [5] which considers the last sheet only partly. Let $S(x)$ denote the number of sheets used in a solution x and c_l the position of the last stage-1 cut of the last sheet. The objective function is then

$$f(x) = \min \left(S(x) - \frac{H - c_l}{H} \right), \tag{1}$$

This refinement allows for a more fine-grained distinction between solutions having an equal number of sheets. When considering real-world applications a cutting pattern yielding one larger waste area is typically preferred as this part can presumably more likely be reused than several smaller ones.

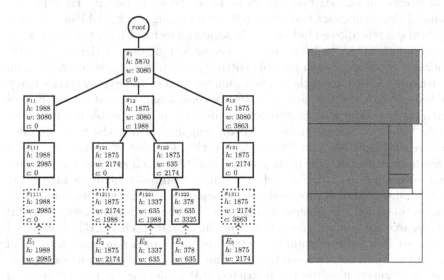

Fig. 1. A cutting tree (left) and the single sheet, s_1, represented by it (right). The leaf nodes $E_1, \ldots, E_5$ represent the actual elements obtained by the applied cutting pattern. The nodes s_{1111} and s_{1211} and s_{1311} are Null-cuts.

2.1 Solution Representation

We represent a solution explicitly by its *cutting tree* (in the literature also referred to as *slicing tree*). Note that a given cutting tree does not necessarily represent a complete solution, but can also be used as a representation of partial solutions or intermediate steps.

The root node represents the whole set of used sheets $S = \{s_1, \ldots, s_n\}$. Each of its children is associated with one sheet (and can be seen as a stage-0 cut). Every further level in the tree corresponds to a guillotine cut of the next stage l that has been applied. Note that the elements from M always appear as leaf nodes at the third level. Whenever an element actually is already finished after the stage-1 or stage-2 cuts, additional so-called *Null-cuts* are introduced to keep this consistent structure.

Each node N in the tree stores the dimensions (h, w) of the represented area, the waste within it and the absolute cut coordinate c on the sheet. If it is a leaf node, the associated element is stored. Figure 1 shows a 3-staged cutting pattern for a single sheet in normal form and the cutting tree representing it.

3 Related Work

The first exact solution approaches to the 2CS have been proposed by Gilmore and Gomory [11], who introduced the "exact two-stage guillotine cutting stock problem", which is a 2-staged 2CS under the additional constraint that all elements packed in one strip of the sheet have the same height. They further introduced the "non-exact two-stage guillotine cutting stock problem" where a final, third stage is allowed but only to separate an element from the waste area. Their approach is the well-known set covering formulation of the problem introducing a variable for each possible cutting pattern of a single sheet. Column generation is used to avoid the explicit enumeration of the exponentially many variables. Oliveira and Ferreira [12], already considering the 3-staged 2CS, presented a faster variant of this approach in which the pricing problem is solved by a greedy heuristic. The exact method is only applied in case the heuristic fails. A more recent approach is by Monaci and Toth [13] whose two-phase algorithm first creates cutting patterns using greedy heuristics which are then the columns in a set covering formulation solved by a Lagrangian-based heuristic algorithm. Alvarez-Valdes et al. [14] proposed a more sophisticated method for solving the pricing problem for the general n-staged 2CS without rotation using GRASP or tabu search for column generation. Puchinger and Raidl [10,15] approached the 3-staged 2CS without rotation restricting the elements in a stack to be of equal width. They use a hierarchy consisting of a greedy construction heuristic, an evolutionary algorithm, a restricted ILP model for the pricing problem and an exact ILP model to generate columns, thus avoiding the computationally expensive uses of the exact method as much as possible.

An important step towards the reduction of the search space was done by Herz [16] who presented the concept of discretization points, excluding the vertical and horizontal coordinates at which no cut can occur in a pattern in normal form. This is also the first definition of so-called "canonical patterns" which correspond to patterns in normal form as defined in section 2. Later Christofides and Whitlock [17] showed a dynamic programming approach to compute them.

Since most of the exact approaches found in the literature are, in fact, hybrid approaches in one way or another, there are also numerous heuristic and metaheuristic approaches for the 2CS that have been studied. In [18] Lodi et al. give a survey on the concept and performance ratios for the prominent greedy heuristics, such as *first-fit*, *first-fit decreasing height* and *finite first-fit*.

4 Variable Neighborhood Search

The *Variable neighborhood search* (VNS) is a metaheuristic which relies on the idea of systematically searching for a better solution in an ordered set of increasingly complex neighborhood structures [19].

Frequently, VNS is combined with *very large(-scale) neighborhood search* (VLNS) techniques. The basic idea of a VLNS in contrast to a "classical" local search approach is to employ a special problem-specific large neighborhood structure $N(x)$, for which an efficient algorithm exists to derive an optimal or good approximate solution [20]. Methods that have been successfully used in VLNS for investigating neighborhoods include shortest path and matching algorithms, dynamic programming, (mixed) integer programming (MIP) and constraint programming. It is not always necessary to solve large neighborhoods to optimality but sometimes also simpler greedy constructive approaches can be applied. These approaches are often used in the context of so-called "ruin-and-recreate" methods, where one iteration consists of destroying randomly chosen or weak parts of an incumbent solution followed by a (usually relatively fast) recreation by a construction heuristic [21].

We follow these considerations in our VNS for the 3-staged 2CS, which is presented in the following.

4.1 Construction Heuristics

As a starting solution for the search the best among the solutions obtained by three different greedy construction heuristics is chosen. These approaches are *3-staged First Fit Decreasing Height with rotations* (3SFFDHR), 3SFFDHR preceded by a matching step (MATCH) and *Fill Strip* (FS), which are summarized in the following.

3SFFDHR. This heuristic is based on the well-known first-fit decreasing height (FFDH) approach for the 2BP. The elements in M are ordered by non-increasing heights h_i; ties are broken randomly. The algorithm iterates through the element list trying to fit the current element in the cutting tree at the first possible position using a post-order traversal. At each level – given the dimensions of the parent and the already existing siblings – it is checked (i) if there is still enough room to accommodate the element, or (ii) if there is enough room for the rotated counterpart of the element. If so, a new sibling is created offering the required height (for a horizontal cut) or width (for a vertical cut) to accommodate the element.

In the worst case the algorithm needs to traverse trough the whole tree for inserting a new element. Therefore the runtime is in $\mathcal{O}(m^2)$.

MATCH. The basic construction principle is the same as for 3SFFDHR. However, one of the disadvantages of the method is that the extent of each area created by a lower stage cut (height for stage-1 and width for stage-2) is fixed based on the element that initialized it.

To increase the space of potential stage-1 heights a preprocessing step is included based on the iterative matching approach by Fritsch and Vornberger [6].

A certain percentage p of the elements in M is chosen randomly and paired into meta-rectangles, s.t. the overall waste is minimal. This is done by first constructing a complete graph $G(V, E)$, where V corresponds to the elements and each edge $(u, v) \in E$ has associated a weight that is inversely proportional to the amount of waste in the bounding rectangle when aligning the elements corresponding to u and v. The meta-rectangles minimizing the overall waste are now determined by calculating a maximum weight matching on this graph.

The resulting meta-rectangles and the remaining elements that were not matched are then packed using the 3SFFDHR heuristic.

As the maximum-weight matching in a graph can be computed in $\mathcal{O}(|V|^3)$, the worst-case runtime of the MATCH construction heuristic is in $\mathcal{O}(m^3)$.

FS. Fill Strip is a strip-based greedy construction heuristic and is an adaptation of the FFFWS heuristic proposed by Puchinger et al. [5]. As for 3SFFDHR the elements are ordered beforehand by non-decreasing height. The algorithm then basically fits the elements according to the same criteria they are fit with 3SFFDHR with the two following modifications:

- Whenever an element does not fit in the current subtree, it is skipped and the algorithm tries to fit the next one from M.
- Once the last element in M is reached, i.e. none of the remaining elements can fit in the current position in the cutting tree, the respective subtree is closed and never reconsidered again in the remaining construction process.
- When a strip is closed, the algorithm continues with the remaining unused elements in M beginning again with the highest one.

In the worst case the algorithm has to restart at the beginning of M for every new strip, hence its worst-case runtime is in $\mathcal{O}(m^2)$.

4.2   Neighborhood Structures and Search

All neighborhood structures we consider follow the principle of a ruin-and-recreate-based VLNS. We choose no "classical" local search neighborhood, since a local search approach based on solutions encoded by e.g. the order of the element list M is prone to suffer from poor locality. Provided no sophisticated decoding algorithm different from simple greedy construction heuristics is employed, a move as basic as a 2-exchange might lead to a rather different cutting pattern. Moreover, evaluating the objective of a solution requires decoding the cutting pattern completely, thereby forfeiting the runtime advantage a local search on a compact encoding usually comes with.

Maximum Waste Ratio. We consider the maximum unused capacity in a used sheet as a secondary fitness value, following an idea that has also been successfully applied for the classical 1BP, cf. [22]. The waste ratio $wr(s)$ of a sheet s is defined as the free area on s (i.e. the area not covered by elements) relative to the total area of s. However, to be consistent with the (main) objective function the last sheet is also considered only partly when determining the maximum. Thus, we have

$$wr(s) = \begin{cases} \dfrac{c_l W - \sum_{i=1 \mid i \in elems(s)}^{m} h_i w_i}{c_l W}, & \text{if } s \text{ is the last sheet} \\[4mm] \dfrac{HW - \sum_{i=1 \mid i \in elems(s)}^{m} h_i w_i}{HW}, & \text{otherwise} \end{cases} \tag{2}$$

where $elems(s)$ is the set of elements that are placed on s and c_l defines the position of the last stage-1 cut of the sheet. The secondary fitness $G(x)$ of a solution x is therefore

$$G(x) = \max_{s \text{ non-empty}} wr(s) \tag{3}$$

$G(x)$ is used as a tie-breaking criterion for solutions with equal objective value to steer the search further towards more compact cutting patterns.

A basic step in our neighborhood search works as follows:

In the incumbent solution one or more subtree(s) of the cutting tree are removed, s.t. the elements associated with the leaves of these sub-trees become free. Using one of the aforementioned construction heuristics they are then reinserted again. Both steps can be efficiently done on the cutting tree representation. Next, we describe the variants for each the ruin and the recreate step in greater detail.

Ruin Subtree. The general parameters for the ruin step are the tree-level λ and either a fixed number of subtrees of the level to be ruined (δ), or a percentage thereof (π). Furthermore, the range with respect to the affected sheets can be controlled. The subtrees may either be chosen randomly from all sheets or restricted to come from the same sheet(s), i.e. they are subtrees of the same level-0 node in the cutting tree. Independent from these settings the last sheet is always affected first, as both the objective function and the secondary fitness capture gradual improvements in terms of the last stage-1 cut of the last sheet. We consider two basic variants:

- *Ruin Random Subtree (RAND)*: Remove the defined number of level-λ subtrees from randomly selected positions in the tree respecting the sheet restrictions.

– *Ruin Max-Waste Subtree (MAX-W)*: Order the available level-λ subtrees by non-increasing waste ratio. Starting with the first one, remove the defined number from the tree respecting the sheet restrictions.

Recreate Cutting Tree. After the ruin step the ruined subtree is normalized, i.e. the remaining subtrees are reordered (and the stored coordinates adapted), s.t. the pattern represented by the tree is in normal form. The removed elements are then sorted again by non-increasing height, shuffled or left unsorted in the order they were removed, before 3SFFDHR, MATCH or FS is used to reinsert them into the tree.

We also consider a next-improvement step function for the possible ruin and recreate combinations. In this work we use a fixed neighborhood order, which is shown in Table 1. Neighborhoods in which the ruin operation is restricted to affecting elements on the same sheet are marked as (f). The first five neighborhoods are chosen for intensification and finding improvements by relatively small changes. The following neighborhoods increasingly perturb the cutting tree by removing random and maximum waste subtrees and reinserting them in various ways. In the last neighborhoods, a special ruin operator is used, the removed subtrees are left intact, s.t. they can be reinserted as a whole (*Soft Remove*).

Table 1. Neighborhoods and their order used in the VNS

k	N_k-Ruin	N_k-Recreate	Step function
1	Random ($\lambda = 3, \delta = 2$)	3SFFDHR Unsorted	Next Imp.
2-3	Random ($\lambda = 2, \delta = k - 1$)	3SFFDHR Unsorted	Next Imp.
4-5	Random ($\lambda = 1, \delta = k - 3$)	3SFFDHR Unsorted	Next Imp.
6-9	Random ($\lambda = 3, \pi = (k - 5) \cdot 0.1$)	3SFFDHR Shuffled	Random
10-13	Random ($\lambda = 3, \pi = (k - 9) \cdot 0.1$)	MATCH Shuffled	Random
14-17	MAX-W ($\lambda = 2, \pi = (k - 13) \cdot 0.1$)	3SFFDHR Shuffled	Random
18-21	MAX-W ($\lambda = 2, \pi = (k - 17) \cdot 0.1$)	MATCH Shuffled	Random
22-25	MAX-W (f) ($\lambda = 2, \pi = (k - 21) \cdot 0.1$)	3SFFDHR Shuffled	Random
26-29	MAX-W (f) ($\lambda = 2, \pi = (k - 25) \cdot 0.1$)	MATCH Shuffled	Random
30-33	Soft ($\lambda = 1, \pi = (k - 29) \cdot 0.1$)	FFDH Shuffled	Random

5   ILP-Based Very Large Neighborhood Search

In the following we introduce a novel compact ILP model for the 2-staged 2CS with variable sheet size. This model is employed in the recreate step of a VLNS neighborhood. More particularly, it is used for optimally packing strips (i.e. defining the stage-2 and stage-3 cuts of the pattern). The remaining problem consists then of packing these strips into sheets.

5.1   An ILP for Packing Strips

Lodi et al. [23] proposed a polynomial-sized ILP model for the 2-staged 2CS. We base our ILP formulation on this model extending it to a model for optimally packing elements in strips of different sizes (in fact, it is a model for the 2-staged 2CS with variable sheet size). The dimensions of the different sheets reflect the dimensions of the stage-1 cuts that can then be placed on the actual sheets.

As defined in section 2 we denote by $t \leq m$ the number of different element types. Let further e_j be the number of elements of type $j \in \{1, \ldots, t\}$ in the demand. The model is now based on the following considerations: We assume that m potential stacks are available. Each of them is associated with a different element a of a certain type i initializing it. Analogously, there are m potential strips, each initialized by a different potential stack a of type k (i.e. a stack initialized by an element of type k).

Furthermore, there are d different dimensions for the strips and each strip, as well as each stack, is of a certain type $l \in \{1, \ldots, d\}$, defined by its dimensions.

We make use of the following observation which is similar to the definition of the normal form of a cutting pattern: For any optimal solution to the problem there exists an equivalent one in which the following conditions hold:

1. The first (topmost) element in each stack is the widest one in the stack.
2. The first (leftmost) element in each strip is the widest one in the strip.

We can further assume an ordering of the element types by nondecreasing width.

Finally, the occurrence of multiple elements of the same type in the demand can be exploited. In a given cutting pattern every permutation of such elements yields another pattern equivalent in structure and quality and can thus be considered symmetrical.

These considerations lead to the following 0/1-variables

$$y_{i_a}^l = \begin{cases} 1 & \text{if the } a\text{-th stack of dimension type } l \text{ is initialized} \\ & \text{by an element of type } i, \\ 0, & \text{otherwise} \end{cases} \tag{4}$$

for $i = 1, \ldots, t; \ l = 1, \ldots, d; \ a = 1, \ldots, m$

$$q_{k_a}^l = \begin{cases} 1 & \text{if the } a\text{-th strip of dimension type } l \text{ is initialized} \\ & \text{by a stack of type } k, \\ 0, & \text{otherwise} \end{cases} \tag{5}$$

for $k = 1, \ldots, t; \ l = 1, \ldots, d; \ a = 1, \ldots, m$
and the positive integer variables

- $x_{i_a j}^l$: The number of elements of type j packed in the a-th stack of dimension type l, initialized by an element of type i.
  for $i = 1, \ldots, t; \ j \geq i; \ l = 1, \ldots, d; \ a = 1, \ldots, m$

– $z^l_{k_a i}$: The number of stacks of type i packed into the a-th strip of dimension type l, initialized by a stack of type k.

for $k = 1, \ldots, t$; $i \geq k$; $l = 1, \ldots, d$; $a = 1, \ldots, m$

Note that the variables inherently satisfy the guillotine cut restriction.

Our variant of the 2-staged 2CS with variable sheet size can now be stated as the following ILP:

$$\min \quad \sum_{l=1}^{d} \sum_{k=1}^{t} \sum_{a=1}^{m} q^l_{k_a} H_l \tag{6}$$

$$\text{s. t.} \quad \sum_{l=1}^{d} \sum_{a=1}^{m} \left(\sum_{i=1}^{j} x^l_{i_a j} + y^l_{j_a} \right) = e_j \quad j = 1, \ldots, t \tag{7}$$

$$\sum_{j=i}^{m} h_j x^l_{i_a j} \leq (H_l - h_i) y^l_{i_a} \quad i = 1, \ldots, t; \ a = 1, \ldots, m; \ l = 1, \ldots, d \tag{8}$$

$$\sum_{a=1}^{m} \left(\sum_{k=1}^{i} z^l_{k_a i} + q^l_{i_a} \right) = \sum_{a=1}^{m} y^l_{i_a} \quad i = 1, \ldots, t; \ l = 1, \ldots, d \tag{9}$$

$$\sum_{i=k}^{m} w_i z^l_{k_a i} \leq (W_l - w_k) q^l_{k_a} \quad k = 1, \ldots, t; \ a = 1, \ldots, m; \ l = 1, \ldots, d \tag{10}$$

$$\sum_{l=1}^{d} y^l_{i_a} \leq 1 \quad i = 1, \ldots, t; \ a = 1, \ldots, m \tag{11}$$

$$\sum_{l=1}^{d} q^l_{k_a} \leq 1 \quad k = 1, \ldots, t; \ a = 1, \ldots, m \tag{12}$$

The objective function (6) minimizes the total height of all used strips. Note that strip a of dimension type l is initialized by an element of type k, iff $q^l_{k_a} = 1$. Equations (7) ensure that each element j is packed exactly e_j times and constraints (8) impose that the height of each used stack does not exceed the respective dimension type's height. Analogously, equations (9) guarantee that each used stack is packed in a used strip while constraints (10) imply that the width of each used strip does not exceed the used dimension type's width. The last two groups of constraints strengthen the model by imposing that each potential stack (11) and each potential strip (12) a can only be used for one specific dimension type.

For the sake of clarity, we keep all variables in the model. However, in an actual implementation we can exclude the following variables, which can never be nonzero. For $1 \leq a \leq m$ and $1 \leq l \leq d$ we can set

$$y_{i_a}^l = 0 \qquad 1 \le i \le t \,|\, h_i > H_l \vee w_i > W_l \tag{13}$$

$$q_{k_a}^l = 0 \qquad 1 \le k \le t \,|\, h_k > H_l \vee w_k > W_l \tag{14}$$

$$x_{i_a j}^l = 0 \qquad 1 \le i,j \le t \,|\, h_i > H_l \vee w_i > W_l \vee h_j > H_l \vee w_j > W_l \vee$$
$$h_i + h_j > H_l \tag{15}$$

$$z_{k_a i}^l = 0 \qquad 1 \le k,i \le t \,|\, h_i > H_l \vee w_i > W_l \vee h_j > H_l \vee w_j > W_l \vee$$
$$w_i + w_j > W_l \tag{16}$$

Constraints (13) and (14) exclude variables for element types that do not fit in the respective dimension type, while (15) and (16) additionally rule out that two element types whose combination exceeds the dimensions can appear together. Finally, we can assume an ordering of the potential stacks and strips in accordance to the order of the element types, i.e.

$$a_1, \ldots, a_{e_1}, a_{(e_1+1)}, \ldots, a_{(e_1+e_2)} \cdots, a_{\sum_{i=1}^{t} e_i}$$

This reflects that each of the stacks (strips) is associated with exactly one element and we can additionally exclude the variables $y_{i_a}^l$ and $x_{i_a j}^l$ ($q_{k_a}^l$ and $z_{k_a i}^l$) according to the range associated with the element type i (k).

Note that the described ILP does not consider rotation of the elements. However, rotation can easily be incorporated by replacing the constraints (7) with

$$\sum_{l=1}^{d} \sum_{a=1}^{m} \left[\left(\sum_{i=1}^{j} x_{i_a j}^l + y_i^l \right) + \left(\sum_{i=1}^{\delta_j} x_{i_a \delta_j}^l + y_{\delta_j}^l \right) \right] = e_j, \quad j = 1, \ldots, 2t; \; j < \delta_j,$$

where δ_j is the index of the rotated variant of type j in the element type order, and by replacing m with $2m$ and t with $2t$ in all the remaining equations, analogously to the variant proposed in [23].

In our ILP-VLNS approach we model the whole cutting tree but fix the variables corresponding to parts not selected by the ruin operator by the respective constants. The resulting strips obtained from solving the ILP are then inserted into the cutting tree, where completely new strips are packed using a FFDH strategy.

This approach is used in two variants within the VNS, see Table 2.

Table 2. Aditional neighborhoods in the ILP-based VNS

k	N_k-Ruin	N_k-Recreate	Step function
34	Random (f) ($\lambda = 3, \pi = 0.33$)	ILP	Random
35	Random (f) ($\lambda = 3, \delta = 0.33$)	ILP	Random

5.2 Determining Strip Dimensions

The remaining open question is how to determine the different strip dimensions, i.e. the different strip heights, that are considered by the model. Clearly, the already used dimensions in the parts of the cutting tree not affected by the ruin operator need to be included. For the additional dimensions there are, however, exponentially many possibilities in the number of element types, even when restricting the choices to the discretization points as proposed in [16].

In our approach this number is reduced during the ILP model generation by restricting the heights to multiples of the heights of available elements. In more detail, a random integer $n \in \{\lceil \frac{h_{\max}}{h_{\min}} \rceil, \ldots, \lceil \frac{H}{h_{\min}} \rceil\}$ is chosen, with $h_{\max}$ and $h_{\min}$ being the largest and the smallest height of all the element types. For each element type i the height dimensions $h_i \cdot k$, with $k \in \{1, \ldots, n\}$ are then generated. As this usually still yields a relatively large number of different heights, a quick evaluation is performed of how promising each of them is. This is done by first shuffling the list of free elements and then letting 3SFFDHR reinsert them into a strip of the given height. The strip heights that yield no more than the average waste ratio are finally chosen.

6 Experimental Results

Our algorithms have been implemented in C++, compiled with GCC version 4.6.3 and executed on a single core of a 3.40 GHz Intel Core i7-3770 with 16 GB RAM. For solving the developed ILP model we have used the general purpose MIP solver CPLEX version 12.6 with default parameter settings and a general time limit of 1000s, as well as a restriction to a single thread. Furthermore, the optimization was stopped as soon as an integer solution having a relative gap of less than or equal to 1% was found.

Computational experiments were performed on the benchmark instances from Berkey and Wang [24] (classes 1 to 6) and Martello and Vigo [25] (classes 7 to 10). Each class consists of 5 subclasses with $m = 20, \ldots, 100$ elements, each of which comprising 10 instances. We compare the basic VNS (VNS SIMPLE), the VNS consisting of the two ILP neighborhoods only (VNS ILP), a combination of both (VNS FULL) and the results obtained by the so far best-performing Branch & Price algorithm from [15] (BPStabEA), which were taken as presented in the respective work. In order to stay comparable to these results rotation of elements is not considered. For VNS SIMPLE, 25 major iterations over all neighborhoods are done, for VNS ILP and VNS FULL the VNS is stopped, when a major iteration does not yield an improvement. Each algorithm (VNS SIMPLE, VNS ILP, VNS FULL, BPStabEA) was applied five times to each problem instance and the average objective value and time spent per instance were determined. These values are then used for computing the average objective $\overline{f(x)}$ and time $\overline{t}$ for the instances of each subclass, which are shown in Table 3. The runtime for each of the experiments was additionally limited to 1000s. Occasionally, this limit was exceeded due to the same but independently measured time limit of 1000s given to CPLEX. The best objective value in each row is printed in bold. In the last rows sums, average and median values over all instances are given.

In general, VNS SIMPLE outperforms VNS ILP and VNS FULL, whereas VNS FULL performs better than VNS ILP for all instance subclasses. We performed one-sided Wilcoxon signed rank tests comparing the objective for each

Table 3. Experimental results. The best objective value in each row is printed in bold.
*Runtimes and objective values for BPStabEA are from [15].

Class	m	VNS SIMPLE $f(x)$	$\bar{t}[s]$	VNS ILP $f(x)$	$\bar{t}[s]$	VNS FULL $f(x)$	$\bar{t}[s]$	BPStabEA $f(x)^*$	$\bar{t}[s]^*$
	20	6.93	1.4	7.06	0.0	**6.90**	0.2	7.2	4.2
	40	**13.34**	10.4	13.47	0.2	13.40	1.4	13.6	201.5
1	60	**20.09**	33.2	20.34	1.0	20.13	5.6	20.1	112.6
	80	27.60	81.5	27.88	1.8	27.71	13.2	**27.5**	68.6
	100	32.24	161.4	32.31	3.2	32.25	22.1	**31.7**	236.9
	20	**0.74**	0.9	0.75	0.0	**0.74**	0.1	1.0	0.1
	40	**1.42**	6.7	1.44	0.5	**1.42**	1.6	2.0	100.5
2	60	**2.12**	22.8	2.14	1.9	**2.12**	4.7	2.7	207.1
	80	2.89	54.6	2.89	63.2	**2.87**	40.3	3.3	228.1
	100	**3.42**	113.9	3.43	89.3	3.43	86.3	4.1	239.7
	20	**4.90**	1.3	5.00	0.1	4.91	0.3	5.4	0.3
	40	**9.56**	9.7	9.83	0.4	9.65	1.7	9.7	6.1
3	60	14.16	33.0	14.46	1.9	14.31	6.1	**14.0**	45.2
	80	19.65	73.1	19.95	4.2	19.69	18.0	**19.2**	166.8
	100	23.23	141.8	23.53	8.0	23.24	44.9	**22.5**	651.4
	20	**0.75**	0.9	**0.75**	0.0	**0.75**	0.1	1.0	0.1
	40	**1.40**	5.8	1.41	2.1	**1.40**	3.8	2.0	100.5
4	60	**2.11**	20.9	2.12	22.7	2.12	21.4	2.6	339.2
	80	2.87	50.9	2.86	499.7	**2.85**	420.2	3.3	321.5
	100	**3.42**	100.2	**3.42**	532.1	**3.42**	618.3	4.0	352.8
	20	**6.30**	1.3	6.35	0.0	6.33	0.2	6.6	0.5
	40	**11.94**	10.1	12.16	0.5	11.97	2.4	12.3	3.0
5	60	18.35	35.1	18.63	2.1	18.46	10.9	**18.3**	10.2
	80	25.34	84.8	25.78	6.5	25.52	25.9	**24.8**	129.7
	100	29.38	164.3	29.75	14.6	29.49	55.9	**28.7**	326.8
	20	**0.66**	0.9	0.67	0.0	0.67	0.1	1.0	0.0
	40	**1.24**	6.4	1.25	1.2	**1.24**	2.4	1.9	400.9
6	60	**1.87**	21.2	**1.87**	45.9	**1.87**	70.2	2.2	118.2
	80	2.53	52.1	2.53	412.6	**2.52**	565.6	3.0	14.0
	100	3.03	107.5	3.03	748.5	**3.02**	818.3	3.6	431.6
	20	**5.16**	1.7	5.25	0.1	5.19	0.2	5.7	0.6
	40	**11.03**	11.7	11.12	0.5	11.05	2.4	11.5	4.8
7	60	**15.89**	42.3	16.09	2.0	15.95	9.6	16.1	22.9
	80	**22.79**	96.9	23.03	5.4	22.84	22.0	23.2	77.8
	100	**27.11**	185.2	27.26	12.2	27.16	47.4	27.1	305.5
	20	**5.69**	1.3	5.92	0.0	5.90	0.1	6.1	1.0
	40	11.42	10.2	11.56	0.7	11.45	1.9	**11.4**	6.7
8	60	16.24	30.9	16.48	6.8	**16.21**	22.2	16.4	30.0
	80	23.00	73.9	23.23	54.5	23.03	64.9	**22.6**	79.5
	100	28.28	143.0	28.40	92.5	28.20	190.6	**28.1**	215.5
	20	**13.86**	2.2	14.04	0.1	13.94	0.2	14.3	0.1
	40	**27.31**	15.8	27.61	0.7	27.49	1.8	27.8	0.3
9	60	**43.29**	55.7	43.64	1.7	43.49	5.6	43.7	0.9
	80	**57.21**	129.1	57.50	7.4	57.27	13.2	57.7	2.6
	100	**69.11**	242.8	69.43	13.3	69.23	29.3	69.5	7.1
	20	**3.96**	1.0	4.06	0.1	3.99	0.2	4.5	0.4
	40	**7.29**	7.9	7.42	0.6	7.33	1.9	7.7	109.6
10	60	**10.14**	24.3	10.41	3.0	10.15	10.1	10.4	461.6
	80	**13.03**	59.1	13.26	8.6	13.07	33.0	13.2	889.1
	100	**16.16**	127.1	16.37	50.4	16.17	100.3	16.4	1000.0
Sum		721.47	2670.0	729.12	2725.3	723.51	3419.2	732.7	8034.1
Average		14.43	53.4	14.58	54.5	14.47	68.4	14.65	160.68
Median		11.22	31.9	11.34	2.0	11.25	9.8	11.45	78.65

instance class using a 95 % confidence interval. For five out of ten instance classes (2,4,6,9 and 10) both VNS SIMPLE and VNS FULL yield significantly better results than obtained by BPStabEA, VNS ILP is significantly better for classes 2,4,6 and 9. VNS FULL does not yield significantly better results than VNS SIMPLE but we observe an average increase in runtime. It can be expected that allowing VNS FULL to run for 25 major iterations would further improve the objective values, however, at the cost of a dramatic increase in the overall runtime.

7   Conclusions

We presented a VNS for the 3-staged 2-dimensional cutting stock problem which uses exclusively "ruin-and-recreate"-based VLNS. In a first straightforward approach greedy construction heuristics were used in the recreate step. We further developed a polynomial-sized ILP as an alternative method for recreating a solution. In fact, this model can also be applied for the 2-staged 2-dimensional cutting stock problem with variable sheet sizes. Experimental results on well-known problem instances show that the hybridization of VNS and VLNS indeed leads to a significant increase in solution quality for half of the instance classes. Using the ILP for recreation of the solution did not significantly increase the performance in comparison to the recreation by construction heuristics.

In future work, we want to improve the ILP-based VLNS in order to reduce the runtime allowing for more searches through larger neighborhoods. To this end we want to develop a more sophisticated approach to determine the dimensions for the strips and strengthen the model itself. Furthermore, it would be interesting to see, if the performance can be improved by inserting the strips with an exact method, e.g. an ILP formulation for the 1BP.

References

1. Lodi, A., Martello, S., Monaci, M.: Two-dimensional packing problems: A survey. European Journal of Operational Research 141(2), 241–252 (2002)
2. Murata, H., Fujiyoshi, K., Nakatake, S., Kajitani, Y.: VLSI module placement based on rectangle-packing by the sequence-pair. IEEE Transactions on Computer-Aided Design of Integrated Circuits and Systems 15(12), 1518–1524 (1996)
3. Wäscher, G., Haußner, H., Schumann, H.: An improved typology of cutting and packing problems. European Journal of Operational Research 183(3), 1109–1130 (2007)
4. Garey, M.R., Johnson, D.S.: "Strong" NP-Completeness Results: Motivation, Examples, and Implications. Journal of the ACM 25(3), 499–508 (1978)
5. Puchinger, J., Raidl, G.R., Koller, G.: Solving a Real-World Glass Cutting Problem. In: Gottlieb, J., Raidl, G.R. (eds.) EvoCOP 2004. LNCS, vol. 3004, pp. 165–176. Springer, Heidelberg (2004)
6. Fritsch, A., Vornberger, O.: Cutting Stock by Iterated Matching. In: Derigs, U., Bachem, A., Drexl, A. (eds.) Operations Research Proceedings 1994. Operations Research Proceedings, vol. 1994, pp. 92–97. Springer, Heidelberg (1995)

7. Cintra, G., Miyazawa, F., Wakabayashi, Y., Xavier, E.: Algorithms for two-dimensional cutting stock and strip packing problems using dynamic programming and column generation. European Journal of Operational Research 191(1), 61–85 (2008)
8. Seiden, S.S., Woeginger, G.J.: The two-dimensional cutting stock problem revisited. Mathematical Programming 102(3), 519–530 (2005)
9. Vanderbeck, F.: A Nested Decomposition Approach to a Three-Stage, Two-Dimensional Cutting-Stock Problem. Management Science 47(6), 864–879 (2001)
10. Puchinger, J., Raidl, G.R.: An Evolutionary Algorithm for Column Generation in Integer Programming: An Effective Approach for 2D Bin Packing. In: Yao, X., et al. (eds.) PPSN 2004. LNCS, vol. 3242, pp. 642–651. Springer, Heidelberg (2004)
11. Gilmore, P.C., Gomory, R.E.: Multistage Cutting Stock Problems of Two and More Dimensions. Operations Research 13(1), 94–120 (1965)
12. Oliveira, J.F., Ferreira, J.S.: A faster variant of the Gilmore and Gomory technique for cutting stock problems. Belgian Journal of Operational Research, Statistics and Computer Science 34(1), 23–38 (1994)
13. Monaci, M., Toth, P.: A Set-Covering-Based Heuristic Approach for Bin-Packing Problems. INFORMS Journal on Computing 18(1), 71–85 (2006)
14. Alvarez-Valdes, R., Parajon, A., Tamarit, J.M.: A computational study of LP-based heuristic algorithms for two-dimensional guillotine cutting stock problems. OR Spectrum 24(2), 179–192 (2002)
15. Puchinger, J., Raidl, G.R.: Models and algorithms for three-stage two-dimensional bin packing. European Journal of Operational Research 183(3), 1304–1327 (2007)
16. Herz, J.C.: Recursive Computational Procedure for Two-dimensional Stock Cutting. IBM Journal of Research and Development 16(5), 462–469 (1972)
17. Christofides, N., Whitlock, C.: An algorithm for two-dimensional cutting problems. Operations Research 25, 30–44 (1977)
18. Lodi, A., Martello, S., Vigo, D.: Recent advances on two-dimensional bin packing problems. Discrete Applied Mathematics 123(13), 379–396 (2002)
19. Hansen, P., Mladenović, N.: Variable Neighborhood Search. In: Glover, F., Kochenberger, G. (eds.) Handbook of Metaheuristics. International Series in Operations Research & Management Science, vol. 57, pp. 145–184. Springer (2003)
20. Ahuja, R.K., Ergun, O., Orlin, J.B., Punnen, A.P.: A survey of very large-scale neighborhood search techniques. Discrete Applied Mathematics 123(1-3), 75–102 (2002)
21. Schrimpf, G., Schneider, J., Stamm-Wilbrandt, H., Dueck, G.: Record Breaking Optimization Results Using the Ruin and Recreate Principle. Journal of Computational Physics 159(2), 139–171 (2000)
22. Benjamin, J., Julstrom, B.A.: Breaking ties with secondary fitness in a genetic algorithm for the bin packing problem. In: Genetic and Evolutionary Computation Conference, pp. 657–664 (2010)
23. Lodi, A., Martello, S., Vigo, D.: Models and bounds for two-dimensional level packing problems. Journal of Combinatorial Optimization 8, 363–379 (2004)
24. Berkey, J.O., Wang, P.Y.: Two-Dimensional Finite Bin-Packing Algorithms. The Journal of the Operational Research Society 38(5), 423–429 (1987)
25. Martello, S., Vigo, D.: Exact Solution of the Two-Dimensional Finite Bin Packing Problem. Management Science 44(3), 388–399 (1998)

Cooperative Parallel Decomposition Guided VNS for Solving Weighted CSP

Abdelkader Ouali[1], Samir Loudni[2], Lakhdar Loukil[1],
Patrice Boizumault[2], and Yahia Lebbah[1]

[1] Université d'Oran, Laboratoire LITIO, BP 1524, El-M'Naouer, 31000 Oran, Algeria
[2] University of Caen, CNRS, UMR 6072 GREYC, 14032 Caen, France

Abstract. Tree decomposition introduced by Robertson and Seymour aims to decompose a problem into clusters constituting an acyclic graph. Recently, Fontaine et al. [8] introduced DGVNS (Decomposition Guided VNS) that uses the *graph of clusters* provided by a tree decomposition to manage the exploration of large neighborhoods. However, for large scale problems, the performance of DGVNS may decrease significantly due to the large number of clusters to be considered sequentially. To overcome this shortcoming we propose CPDGVNS (Cooperative Parallel DGVNS) in which the clusters are explored in parallel through an asynchronous master-slave architecture. Experiments performed on real life instances show the appropriateness and the efficiency of our approach.

Keywords: Tree decomposition, Weighted CSP, Parallelization, Meta-heuristics, Variable Neighborhood Search (VNS), Master-Slave architecture.

1 Introduction

Tree decomposition introduced by Robertson and Seymour [17] aims to decompose a problem into subproblems (called clusters) constituting an acyclic graph. Each *cluster* corresponds to a subset of variables that are strongly connected in the initial graph. Once decomposed, the solving time of the initial problem can be bounded by an exponent of its *width*, which is the size of the largest cluster in the tree (minus 1). This nice property explains why tree decomposition received a great interest in various domains: for checking satisfiability in SAT [16], for solving Constraint Satisfaction Problem (CSP) [6], in Bayesian or probabilistic networks [13], in relational databases [9], for constraint optimization [5,18]. All these proposals exploit tree decomposition for complete search methods.

More recently, Fontaine et al. [8] investigated the incorporation of tree decomposition within *Variable Neighborhood Search* (VNS) [14]. They proposed DGVNS (Decomposition Guided VNS) that exploits the *graph of clusters* provided by a tree decomposition of the constraints graph of the problem to build neighborhood structures. However, as happens with other meta-heuristics, the main limitation with DGVNS is that, for very large scale problems, its performance may decrease significantly due to the large number of clusters to be considered.

M.J. Blesa, C. Blum, and S. Voß (Eds.): HM 2014, LNCS 8457, pp. 100–114, 2014.
© Springer International Publishing Switzerland 2014

In this paper, we propose a first parallelization strategy for DGVNS called CPDGVNS (Cooperative Parallel DGVNS) that consists simply in exploring all of the clusters in parallel. CPDGVNS follows a master-slave architecture, where the master process keeps, updates, and communicates the current overall best solution, while the slave processes manage the exploration of individual clusters. The individual processes cooperate by asynchronously exchanging information about the best solutions computed so far. That ensures independence of the individual slave processes and allows starting with various initial solutions, thus enabling more diversification.

Experiments performed on real life instances (RLFAP, SPOT5 and tagSNP) show that, compared with DGVNS, CPDGVNS provides significant improvements, both in terms of success rate and CPU times, on RLFAP instances (particularly Scen08) and on large tagSNP instances. To the best of our knowledge, our proposal constitutes the first attempt to use tree decomposition to efficiently parallelize the exploration of large neighborhoods in VNS.

Section 2 introduces the context. Section 3 presents how to exploit tree decomposition to efficiently parallelize the exploration of large neighborhoods within VNS. Section 4 presents the problem instances we used for our experiments. Section 5 is devoted to experimentations. Finally, we conclude and draw some perspectives.

2 Context and Definitions

First, we recall the definition of Weighted CSP, the framework we have retained for modeling all problems considered for our experiments (see Section 4). Then, we present the MCS[1] tree decomposition method that relies on the concept of graph triangulation. Finally, we detail the DGVNS method.

2.1 Weighted CSP

A weighted CSP (WCSP) [11] is a generic framework used to model and solve constrained optimization problems which allows to deal with over-constrained problems. They have been successfully applied to resource allocation [3], scheduling [2], bio-informatics [19] and probabilistic reasoning [15].

A WCSP is a pair (X, W) where $X = \{x_1, \ldots, x_n\}$ is a set of n variables (with a maximum domain size d) and W is a set of e cost functions. Each variable $x_i \in X$ has a finite domain D_i of values that can be assigned to it. A value a in D_i is denoted (x_i, a). For a set of variables $S \subseteq X$, D^S denotes the cartesian product of the domains of the variables in S. A *complete* assignment $t = (a_1, \ldots, a_n)$ is an assignment of all variables; on the contrary, it will be called a *partial* assignment. For a given complete assignment t, $t[S]$ denotes the projection of t over S. A cost function $w_S \in W$, with scope $S \subseteq X$, is a function $w_S : D^S \mapsto [0, k_\top]$ where, $k_\top$ is a maximum integer cost (finite or not) used to represent forbidden assignments

---

[1] Maximum Cardinality Search.

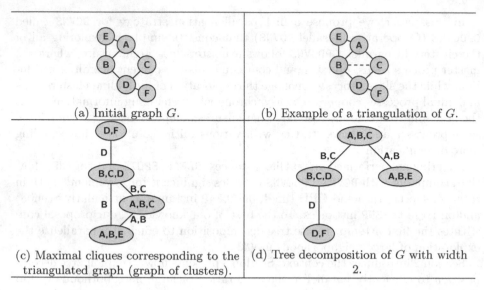

(a) Initial graph G.

(b) Example of a triangulation of G.

(c) Maximal cliques corresponding to the triangulated graph (graph of clusters).

(d) Tree decomposition of G with width 2.

Fig. 1. Steps for computing a tree decomposition of a graph G

(expressing hard constraints). Costs are combined using the bounded addition defined by $\alpha \oplus \beta = \min(k_T, \alpha + \beta)$. Solving a WCSP consists in finding a complete assignment t minimizing $\oplus_{w_S \in W} w_S(t[S])$.

2.2 Tree Decomposition

The constraints graph of a WCSP is a graph $G=(X,E)$ with one vertex for each variable and one edge (u,v) for every cost function $w_S \in W$, such that $u,v \in S$.

Definition 1. *A tree decomposition [17] of $G=(X,E)$ is a pair (C_T,T) where $T = (I,A)$ is a tree with nodes set I and edges set A, and $C_T = \{C_i \mid i \in I\}$ is a family of subsets of X (called clusters) such that: (i) $\cup_{i \in I} C_i = X$, (ii) $\forall (u,v) \in E, \exists C_i \in C_T$ s.t. $u, v \in C_i$, (iii) $\forall i,j,k \in I$, if j is on the path from i to k in T, then $C_i \cap C_k \subseteq C_j$.*

The intersection of two clusters C_i and C_j is called a *separator*, and noted $sep(C_i,C_j)$. Two clusters C_i and C_j are *adjacent* if $sep(C_i,C_j) \neq \emptyset$. The neighborhood of C_i, denoted $neighbor(C_i)$, is the set of clusters C_j that are adjacent to C_i.

Definition 2. *A graph of clusters for a tree decomposition (C_T,T) is an undirected graph $G_T = (C_T, E_T)$ that has a vertex for each cluster $C_i \in C_T$, and there is an edge $(C_i,C_j) \in E_T$ when $sep(C_i,C_j) \neq \emptyset$. The edges are labeled by the shared variables.*

There has been a lot of work on tree decompositions. Usually, the problem considered is to produce a decomposition with a minimum treewidth, an

---

Algorithm 1. DGVNS

Require: The constraint graph (X,W), initial number of variables to unassign k_{init}, maximum number of variables to unassign k_{max}, discrepancy δ_{max} for LDS.
1: let G be the constraints graph of (X, W)
2: let (C_T, T) be a tree decomposition of G
3: let $C_T = \{C_1, C_2, ..., C_p\}$
4: $S \leftarrow$ GENRANDOMSOL()
5: $k \leftarrow k_{init}$, $i \leftarrow 1$
6: **while** $(k < k_{max}) \wedge$ (not TimeOut) **do**
7:      $Cand \leftarrow$ COMPLETECLUSTER(C_i, k)
8:      $X_{un} \leftarrow$ HNEIGHBORHOOD($Cand, k, S$)
9:      $\mathcal{A} \leftarrow S \backslash \{(x_i, a) \mid x_i \in \mathcal{X}_{un}\}$
10:     $S' \leftarrow$ LDS+CP($\mathcal{A}, X_{un}, \delta_{max}, f(S), S$)
11:     NEIGHBORHOODCHANGE(S, S', k, i)
12: **end while**
13: **return** S

Algorithm 2. Neighborhood Change

1: **procedure** NEIGHBORHOODCHANGE(S, S', k, i)
2:      **if** $f(S') < f(S)$ **then**
3:          $S \leftarrow S'$
4:          $k \leftarrow k_{init}$, $i \leftarrow succ(i)$
5:      **else**
6:          $k \leftarrow k + 1$, $i \leftarrow succ(i)$
7:      **end if**
8: **end procedure**

NP- hard problem [1]. Approximate tree decompositions using *triangulation* of a given graph are often exploited. We used *Maximum Cardinality Search* (MCS) [20] heuristic aimed at the production of tree decompositions with small treewidth.

Fig. 1 depicts the three steps for computing a tree decomposition of a graph G (see Part a). First, triangulation is performed on G by adding edge BC (see Part b). Then, maximal cliques in the chordal graph are determined in order to build the graph of clusters (see Part c). Finally, tree decomposition is achieved (see Part d).

2.3   Decomposition Guided VNS (DGVNS)

DGVNS (Decomposition Guided VNS) [8] extends the Variable Neighborhood Search (VNS) method [14], by exploiting the graph of clusters in order to guide the exploration of large neighborhoods. Neighborhoods are obtained by unfixing a part of the current solution according to a neighborhood heuristic. Then the exploration of the search space, related to the unfixed part of the current solution, is performed by a partial tree search LDS (*Limited Discrepancy Search*, [10]) with Constraint Propagation (CP).

Definition 3 (Neighborhood Structure $N_{k,i}$). *Let G be a constraint graph and $G_T{=}(\mathcal{C}_T,E_T)$ its associated graph of clusters. Let $C_i \in \mathcal{C}_T$ be a cluster of G_T and k the neighborhood dimension. $N_{k,i}$ denotes the set of all subsets of k variables from C_i.*

Algorithm 1 depicts the pseudo-code of DGVNS. It starts from a tree decomposition of G (line 2) and from an initial solution S which is randomly generated (line 4). To favor moves on regions that are closely linked, DGVNS uses neighborhood structures $N_{k,i}$ (see Definition 3). Indeed, the concept of cluster embodies this criterion, because of its size (smaller than the original problem), and by the strong connection of the variables it contains. Thus, the set of candidate variables $Cand$ to be unassigned are selected from cluster C_i. If $(k > |C_i|)$, then we complete $Cand$ by adding the clusters C_j adjacent to C_i in order to take into account the topology of the graph of clusters. This treatment is achieved by function COMPLETECLUSTER(C_i, k) (line 7). A subset of k variables X_{un} is randomly selected in $Cand$ among conflicted ones[2] by the neighborhood heuristic HNEIGHBORHOOD$(Cand, k, S)$ (line 8). A partial assignment $\mathcal{A}$ is generated from the current solution S by unassigning the k selected variables; the $(n - k)$ non-selected variables keep their current value in S (line 9). Then, unassigned variables are rebuilt by a partial tree search (LDS) combined with Constraint Propagation (CP) (line 10). The search stops when the maximal dimension size allowed or the $TimeOut$ is reached (line 6).

To achieve a better diversification, DGVNS considers successively all the C_i. This treatment is achieved by procedure NEIGHBORHOODCHANGE(S, S', k, i) (line 11). Let p be the total number of clusters, $succ$ a successor function[3], and $N_{k,i}$ the current neighborhood structure. Initially, k is set to k_{init} (line 5). In neighborhood change strategy (Algorithm 2), if LDS+CP finds a solution of better quality (line 2), then S' becomes the current solution (line 3), k is reset to k_{init} and the next cluster is considered (line 4). Otherwise, we look for improvements in $N_{(k+1),succ(i)}$ (neighborhood structure where $(k+1)$ variables of $Cand$ will be unassigned (line 6)).

3   Parallel DGVNS

First, we discuss and motivate the architecture we considered for our approach. Then, we detail the role played by the master and the slave processes.

3.1   Asynchronous Master-Slave Architecture for CPDGVNS

The main motivation behind parallelism is to improve the performance of the algorithms with respect to the computational time and the solution quality. Several studies on parallel meta-heuristics confirm this trend. The parallel meta-heuristic

---

[2] A variable is said to be conflicted if it occurs in at least one unsatisfied constraint.
[3] if $i < p$ then $succ(i) = i + 1$ else $succ(p) = 1$.

literature also indicates that while multiple independent search strategies provide good results, they are generally outperformed by well-designed *cooperative asynchronous* search strategies (see [4] for more details). That is why we have adopted this approach.

In order to explore a larger part of the solution space, we propose the Cooperative Parallel DGVNS (CPDGVNS), that explores in parallel all of clusters provided by a tree decomposition. CPDGVNS follows a *master-slave* architecture. The main tasks performed by processes in CPDGVNS can be summarized as follows:

- **master process:**
    - Sends to each slave process the initial solution and the cluster C_i to be explored;
    - Following a communication from a slave process, updates the best overall solution;
    - Verifies the stopping condition and relaunches the available slave process starting from the best overall solution in $N_{k_{init}, succ(i)}$.
- **each slave process:**
    - Receives the initial solution and manages the exploration of the assigned cluster;
    - Communicates its solution to the master.

The slave processes cooperate by exchanging information about the best solutions computed so far. They communicate exclusively with a master process. Solution updates and communications are performed *asynchronously*. This makes this approach advantageous over the synchronous one as it allows starting with various initial solutions during the successive resolutions of the clusters, thus enabling more diversification.

3.2 Master Algorithm

Let $C_T = \{C_1, ..., C_p\}$ be the set of clusters and n_{sl} the number of slave processes used. In our approach, to fully benefit from the parallelization, we fixed the number of processes n_{sl}[4] to the number of clusters. Thus, slave processes are ranked from 1 to n_{sl}, while the master process is ranked zero. CPDGVNS starts from a tree decomposition of G and from an initial solution S which is randomly generated (lines 3-5, Algorithm 3). It proceeds in three steps:

Initialization Step. The master initiates the search by launching in parallel the execution of n_{sl} slave processes (lines 7-11, Algorithm 3). This is done by sending to each slave r the same initial solution, the corresponding cluster C_i from C_T, and the values of parameters k_{init}, k_{max}, and δ_{max} (line 9). List C_T is managed as a FIFO strategy in order to ensure that every cluster is processed by a unique slave process (line 10). Initially, the value of k_{max} is set to the size of the cluster assigned to each slave process (line 8). This restricts the choice of variables to be unassigned only to variables of this cluster.

---

[4] Conceptually, the number of slave processes is equal to the number of clusters. In practice, if the number of available physical cores is less than the number of clusters, the same core will be used to process different clusters.

Algorithm 3. Master process

1: **function** CPDGVNS($X, W, k_{init}, \delta_{max}, n_{sl}$)
2:     let G be the constraints graph of (X, W)
3:     let (C_T, T) be a tree decomposition of G
4:     let $C_T = \{C_1, C_2, ..., C_p\}$
5:     $S \leftarrow$ GENRANDOMSOL()
6:     $i \leftarrow 1$
7:     **for each** slave $r = 1, \ldots, n_{sl}$ **do**
8:         $k_{max} \leftarrow |C_i|$
9:         SEND($r, i, k_{init}, k_{max}, \delta_{max}, S$)
10:         $i \leftarrow succ(i)$
11:     **end for**
12:     $Finished \leftarrow 0, adj \leftarrow 0$
13:     **while** $(Finished < n_{sl})$ **do**
14:         RECEIVE(r, S'_r)
15:         **if** $(f(S'_r) < f(S))$ **then**
16:             $S \leftarrow S'_r, adj \leftarrow 0$
17:             $i \leftarrow succ(i), k_{max} \leftarrow |C_i|$
18:         **else**
19:             $i \leftarrow succ(i), k_{max} \leftarrow |C_i|$
20:             $adj \leftarrow adj + 1$
21:             **for** $j = 1, \ldots, adj$ **and** $adj \leq |neighbor(C_i)|$ **do**
22:                 Select the jth cluster C_j from $neighbor(C_i)$
23:                 $k_{max} \leftarrow k_{max} + |C_j|$
24:             **end for**
25:         **end if**
26:         **if** (not global_TimeOut) **then**
27:             SEND($r, i, k_{init}, k_{max}, \delta_{max}, S$)
28:         **else**
29:             $Finished + +$
30:         **end if**
31:     **end while**
32:     **return** S
33: **end function**

Updating Step. The master waits for the best solution found by each slave process (lines 14-25, Algorithm 3). Let S'_r be the new solution communicated by the slave process r to the master. If S'_r is of better quality than S (line 15), S'_r becomes the best overall solution (line 16), the next cluster C_i is considered and k_{max} is reset to $|C_i|$ (line 17). Otherwise, we look for improvements in the next cluster C_i (line 19) and we enlarge the set of candidate variables to be unassigned by adding clusters C_j adjacent to C_i. This treatment is achieved by increasing the number of adjacent clusters adj to be considered (line 20) and the value of k_{max} accordingly (lines 21-23). This is done each time the slave does not succeed to improve the best overall solution.

First, diversification performed by moving from cluster C_i to cluster $C_{succ(i)}$ is necessary. Experiments we performed have shown that remaining in the same

---

Algorithm 4. Slave process r

Require: Tree decomposition (C_T, T)
1: RECEIVE$(0, i, k_{init}, k_{max}, \delta_{max}, S)$
2: $S_r \leftarrow S$
3: $k \leftarrow k_{init}$
4: **while** $(k < k_{max}) \wedge$(not local_TimeOut) **do**
5:    $Cand \leftarrow$ COMPLETECLUSTER(C_i, k)
6:    $X_{un} \leftarrow$ HNEIGHBORHOOD$(Cand, k, S_r)$
7:    $\mathcal{A} \leftarrow S_r \backslash \{(x_i, a) \mid x_i \in \mathcal{X}_{un}\}$
8:    $S_r' \leftarrow$ LDS+CP$(\mathcal{A}, X_{un}, \delta_{max}, f(S_r), S_r)$
9:    NEIGHBORHOODCHANGE(S_r, S_r', k)
10: **end while**
11: SEND$(0, S_r)$
12: **procedure** NEIGHBORHOODCHANGE(S, S', k)
13:    **if** $f(S') < f(S)$ **then**
14:       $S \leftarrow S'$
15:       $k \leftarrow k_{init}$
16:    **else**
17:       $k \leftarrow k + 1$
18:    **end if**
19: **end procedure**

---

cluster leads to lower improvements: selecting a new cluster enables to improve the quality of the solution by visiting new parts of the search space. Second, when a local minimum is found in the current neighborhood, increasing the value of k_{max} will also provide some diversification by enlarging the neighborhood size.

Intensification Step. The aim of this step is to use the best local optimum found by the processes to improve the intensification of the search (lines 26-30, Algorithm 3). Thus, if the *global_TimeOut* is not reached, the search is continued by re-launching the slave process r starting from the best available overall solution (line 27). Otherwise, it is stopped (line 32). The whole solving process terminates when all of the slave processes finish (line 13).

3.3   Slave Algorithm

The aim of slave processes is to improve the solution of the master by unfixing k variables of this solution in the neighborhood structure $N_{k,i}$, where C_i is the cluster where the variables will be selected from, and rebuilding them using LDS+CP.

Algorithm 4 depicts the pseudo-code of the slave process r. It requires the tree decomposition (C_T, T) of G. It receives from the master, the index of the assigned cluster, the values of parameters k_{init}, k_{max}, the value of discrepancy δ_{max} for LDS+CP, and the initial solution S (line 1). As for DGVNS, the set of candidate variables $Cand$ to be unassigned are selected from cluster C_i. If $(k > |C_i|)$ and $(k_{max} > |C_i|)$ (see treatment achieved by lines 21-23, Algorithm 3), then we complete $Cand$ by adding the clusters C_j adjacent to C_i. This treatment is achieved by function COMPLETECLUSTER(C_i, k) (line 5). A subset of k variables X_{un} is randomly selected in $Cand$ among conflicted ones by the neighborhood heuristic HNEIGHBORHOOD$(Cand, k, S_r)$ (line 6). A partial assignment $\mathcal{A}$ is generated from the current solution S_r by unassigning the k selected variables; the $(n - k)$ non-selected variables keep their current value in S (line 7). Then, unassigned variables are rebuilt using LDS+CP (line 8).

If LDS+CP finds a solution of better quality S'_r in the neighborhood of S_r (line 13), then S'_r becomes the current solution (line 14) and k is reset to k_{init} (line 15). Otherwise, contrary to DGVNS, the slave process looks for improvements in the neighborhood structure where $(k + 1)$ variables of X will be unassigned (line 17). This treatment is achieved by procedure NEIGHBORHOODCHANGE(S_r, S'_r, k) (line 9). The search stops when it reaches the maximal number of variables to be unassigned k_{max} or the $local\_TimeOut$ (line 4).

4   Benchmark Problems

Experiments have been performed on instances of three different problems modeled as Cost Function Network (CFN) (see Section 2.1).

RLFAP instances: The CELAR (Centre d'Electronique de l'Armement) has made available a set of instances for the Radio Link Frequency Assignment Problem (RLFAP) [3]. They consist in assigning a limited number of frequencies to a set of radio links defined between pairs of sites, in order to minimize interferences due to the re-use of frequencies. We report experiments on the most difficult instances: Scen06, Scen07 and Scen08.

SPOT5 instances: The daily management of an earth observation satellite such as SPOT5 consists in selecting a subset of candidate photographs to fit physical limitations and maximize the importance of the selected photographs [2]. We report experiments on six instances from those without hard capacity constraint.

tagSNP instances: A Single Nucleotide Polymorphism (SNP) is a DNA sequence variation occurring when a single nucleotide - A, T, C or G - in the genome differs between members of a biological species or paired chromosomes in an individual [7]. SNPs act as biological markers that may help predict risk of developing particular diseases. The tagSNP problem consists in selecting a small subset of SNPs, called tagSNPs, that captures most of the genetic information. This problem is known to be very hard to solve, due to its close relation to the *set covering problem* (NP-Hard) [18]. We report experiments on twelve challenging instances derived from human chromosome-1-data[5] with r_0=0.5 (up to n=1550 variables with maximum domain size d ranging from 30 to 266, and up to e=250,000 cost functions). Six instances are medium-sized, and the six other ones are large-sized.

---

[5] http://www.costfunction.org/benchmark

5   Experiments

We compare CPDGVNS with DGVNS, on RLFAP, SPOT5 and tagSNP instances. Experiments we performed clearly show:

- The relevance of parallelizing the exploration of the graph of clusters.
- Compared with DGVNS, CPDGVNS provides significant improvements, both in terms of success rate and CPU times, on RLFAP instances (particularly Scen08) and on large tagSNP instances.

5.1   Experimental Protocol

To compare CPDGVNS to DGVNS, we have taken the same parameters as those described in [8]. The value of discrepancy for LDS is set to 3 which is the best value found on RLFAP instances (see [12]). k_{min} and k_{max} are respectively set to 4 and n (the total number of variables) so that all variables of the problem will be covered, and $global\_TimeOut$ fixed to 3600 seconds. For CPDGVNS, the number of processes n_{sl} is fixed to $|C_T|$ which is the number of clusters of the tree decomposition.

The experimentations have been carried out on the high performance computing center of the university of Oran[6]. A set of 50 runs per instance has been performed. All search strategies have been implemented in C++ using the library toulbar2[7]. The parallelization has been done within MPI (Message Passing Interface) environment [8]. For instance, SEND and RECEIVE synchronization routines correspond respectively to MPI_SEND and MPI_RECV procedures of the MPI library.

To evaluate the impact of our parallel strategy, we compare the quality of solutions obtained by DGVNS and CPDGVNS by considering the computation time. To implement this purpose, we fixed the $local\_TimeOut$ allocated to each slave process to $global\_TimeOut/n_{sl}$. For each instance and each method, we report:

1. the number of successful runs; a run is successful if the optimum is reached.
2. the average CPU time (in seconds) for the successful runs,
3. the average cost of the final best solution over the 50 runs,

5.2   Contribution of the Parallelization

RLFAP instances. CPDGVNS clearly outperforms DGVNS on RLFAP instances, both in terms of success rates and CPU times (see Table 1). CPDGVNS reaches the optimum with success rate of 100% on all the instances. For Scen06, CPDGVNS improves the success rate about 10% (from 90% to 100%) and CPDGVNS is 9.4 times faster than DGVNS. For Scen07, the two methods get the same success rates, but CPDGVNS is more faster. For Scen08, one of the most challenging instances,

---

[6] http://www.univ-oran.dz/uci/index.html
[7] http://carlit.toulouse.inra.fr/cgi-bin/awki.cgi/SoftCSP
[8] http://www.mcs.anl.gov/research/projects/mpi/

Table 1. Comparing CPDGVNS and DGVNS on RLFAP instances

Instance	Method	Succ.	Time	Avg.
Scen06 $S^* = 3,389$	CPDGVNS($n_{sl} = 12$)	**50/50**	**5.28**	**3,389**
$n = 100,\ d = 44,\ e = 1,222$	DGVNS	45/50	49.70	3390
Scen07 $S^* = 343,592$	CPDGVNS($n_{sl} = 19$)	**50/50**	**221.07**	**343,592**
$n = 200,\ d = 44,\ e = 2,665$	DGVNS	50/50	344.48	343,592
Scen08 $S^* = 262$	CPDGVNS($n_{sl} = 46$)	**50/50**	**371.57**	**262**
$n = 458,\ d = 44,\ e = 5,286$	DGVNS	15/50	826.26	273

Table 2. Comparing CPDGVNS and DGVNS on SPOT5 instances

Instance	Method	Succ.	Time	Avg.
#408 $S^* = 6,228$	CPDGVNS($n_{sl} = 9$)	**50/50**	**6.49**	**6,228**
$n = 200, e = 2,232$	DGVNS	28/50	2.06	6,228
#412 $S^* = 32.381$	CPDGVNS($n_{sl} = 9$)	**50/50**	**16.05**	**32,381**
$n = 300, e = 4,348$	DGVNS	27/50	7.67	32,381
#414 $S^* = 38,478$	CPDGVNS($n_{sl} = 14$)	**50/50**	**66.95**	**38,478**
$n = 364, e = 10,108$	DGVNS	19/50	54.44	38,479
#505 $S^* = 21,253$	CPDGVNS($n_{sl} = 12$)	**40/50**	**6.07**	**21,253**
$n = 240, e = 2,242$	DGVNS	16/50	3.67	21,254
#507 $S^* = 27,390$	CPDGVNS($n_{sl} = 11$)	**47/50**	**33.16**	27,390
$n = 311, e = 5,732$	DGVNS	25/50	15.72	27,390
#509 $S^* = 36,446$	CPDGVNS($n_{sl} = 13$)	**50/50**	**63.14**	**36,446**
$n = 348, e = 8,624$	DGVNS	29/50	38.49	36,447

the trend is greatly amplified: CPDGVNS improves very significantly the success rate about 70% (from 30% to 100%) and obtains solutions with a *mean deviation* (percentage deviation from the optimum) of 0% above the optimum against 4.2% for DGVNS. Moreover, CPDGVNS is 2.22 times faster than DGVNS.

SPOT5 Instances. Table 2 confirms the robustness of CPDGVNS on SPOT5 instances, where CPDGVNS reaches the optimum with success rate of 100% on the instances (#408, #412, #414, #509), and DGVNS gets on average a success rate about 51.5%. For instance #505, CPDGVNS improves the success rate about 48%, while for #507 it improves the success about 44%.

Finally, Table 2 suggests that DGVNS is (on average) faster than CPDGVNS. This is clearly due to the fact that DGVNS gets significant less successful runs compared with CPDGVNS, thus impacting the average CPU-times. However, if we consider unsuccessful runs, CPDGVNS clearly outperforms DGVNS.

tagSNP Instances. Table 3 compares CPDGVNS and DGVNS on tagSNP instances. For *medium-sized instances*, CPDGVNS clearly dominates DGVNS in terms of CPU times. CPDGVNS reaches the optimum for each of the 50 runs (i.e. success rates of 100%). DGVNS gets the same success rates (except for instance #8956), but CPDGVNS is on average 3.39 times faster. For the instance #8956, CPDGVNS improves the success rate about 4%.

Table 3. Comparing CPDGVNS and DGVNS on tagSNP instances

Instance	Method	Succ.	Time	Avg.
#3792, $n = 528$, $d = 59$,	CPDGVNS($n_{sl} = 70$)	**50/50**	**19.57**	**6,359,805**
$e = 12,084$, $S^* = 6,359,805$	DGVNS	50/50	90.59	6,359,805
#4449, $n = 464$, $d = 64$,	CPDGVNS($n_{sl} = 56$)	**50/50**	**16.61**	**5,094,256**
$e = 12,540$, $S^* = 5,094,256$	DGVNS	50/50	36.81	5,094,256
#8956, $n = 486$, $d = 106$,	CPDGVNS($n_{sl} = 54$)	**50/50**	**19.24**	**6,660,308**
$e = 20,832$, $S^* = 6,660,308$	DGVNS	48/50	85.99	6,666,309
#9319, $n = 562$, $d = 58$,	CPDGVNS($n_{sl} = 62$)	**50/50**	**10.05**	**6,477,229**
$e = 14,811$, $S^* = 6,477,229$	DGVNS	50/50	33.23	6,477,229
#16421, $n = 404$, $d = 75$,	CPDGVNS($n_{sl} = 35$)	**50/50**	**14.15**	**3,436,849**
$e = 12,138$, $S^* = 3,436,849$	DGVNS	50/50	48,19	3,436,849
#16706, $n = 438$, $d = 30$,	CPDGVNS($n_{sl} = 49$)	**50/50**	**5.13**	**2,632,310**
$e = 6,321$, $S^* = 2,632,310$	DGVNS	50/50	11.95	2,632,310
#6858, $n = 992$, $d = 260$,	CPDGVNS($n_{sl} = 105$)	**40/50**	**308.93**	**20,833,413**
$e = 103,056$, $S^* = 20,162,249$	DGVNS	33/50	2788.92	20,565,101
#9150, $n = 1,352$, $d = 121$,	CPDGVNS($n_{sl} = 120$)	**50/50**	**260.07**	**43,301,891**
$e = 44,217$, $S^* = 43,301,891$	DGVNS	27/50	2660.00	43,497,252
#14007, $n = 1554$, $d = 195$,	CPDGVNS($n_{sl} = 31$)	**50/50**	**665.54**	**50,290,563**
$e = 54,753$, $S^* = 50,290,563$	DGVNS	19/50	2,523.38	50,913,924
#10442, $n = 908$, $d = 76$,	CPDGVNS($n_{sl} = 25$)	**50/50**	**168.59**	**21,591,913**
$e = 28,554$, $S^* = 21,591,913$	DGVNS	50/50	228.50	21,591,913
#14226, $n = 1,058$, $d = 95$,	CPDGVNS($n_{sl} = 94$)	**50/50**	**159.18**	**25,665,437**
$e = 36,801$, $S^* = 25,665,437$	DGVNS	50/50	295.78	25,665,437
#17034, $n = 1,142$, $d = 123$,	CPDGVNS($n_{sl} = 120$)	**50/50**	**166.65**	**38,318,224**
$e = 47,967$, $S^* = 38,318,224$	DGVNS	50/50	565.06	38,318,224

For *large instances*, CPDGVNS clearly outperforms DGVNS both in terms of success rates and CPU times, particularly on the three instances #6858, #9150 and #14007. CPDGVNS improves the success rate about 14% on instance #6858, 46% on instance #9150, and 62% on instance #14007. Moreover, for these three instances, CPDGVNS is up to 10 faster than DGVNS (it is on average 7 times faster). For the other instances, CPDGVNS is on average 2.2 times faster than DGVNS.

Performance Profile. We have selected four instances to describe the *mean performance profiles* of the evolution of the average solution quality over time for the 50 runs of both CPDGVNS and DGVNS on the instances of RLFAP and tagSNP. Two instances are from RLFAP (Scen06 and Scen08), and two other large instances from tagSNP (#14007 and #17034).

Fig. 2 compares the performance profiles of both CPDGVNS and DGVNS on Scen06 and Scen08 respectively. From an anytime point of view, CPDGVNS clearly outperforms DGVNS especially on Scen08, where the curve of CPDGVNS shows a significant steep initial slope. This remark is also confirmed on the two large tagSNP instances (#14007 and #17034, Fig. 3). For instance Scen08, on average, CPDGVNS intensifies the search around the optimum after 60 seconds of

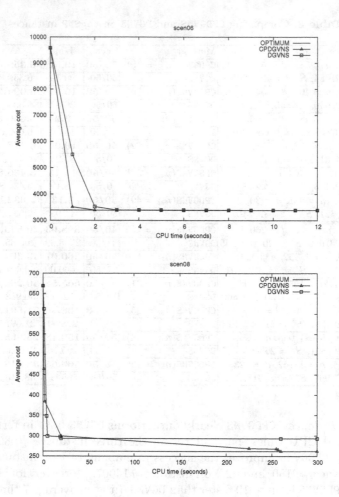

Fig. 2. Profile performance of CPDGVNS and DGVNS on RLFAP instances: Scen06 and Scen08

computation time, while DGVNS takes about 120 seconds to get close to the optimum. Indeed, experiments showed that CPDGVNS escape easily from the local optimum compared to DGVNS due to the multiple search on different clusters. This confirms the contribution of parallelism to compromise between diversification and intensification.

Synthesis. These experiments clearly demonstrate the efficiency of CPDGVNS. Compared with DGVNS, CPDGVNS provides significant improvements, both in terms of success rate and CPU times, on RLFAP instances (particularly Scen08) and on large tagSNP instances.

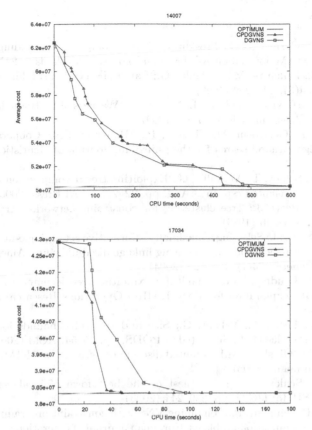

Fig. 3. Profile performance of CPDGVNS and DGVNS on tagSNP instances: #14007 and #17034

6 Conclusions

In this paper, we have proposed a first parallelization strategy for DGVNS called CPDGVNS (Cooperative Parallel DGVNS), that explores in parallel all of clusters provided by a tree decomposition. CPDGVNS follows a master-slave architecture, where the master process keeps, updates, and communicates the current overall best solution, while the slave processes manage the exploration of individual clusters. The individual processes cooperate by asynchronously exchanging information about the best solutions computed so far, thus enabling more diversification. Experimental results show that CPDGVNS provides significant improvements, both in terms of success rate and CPU times.

We are currently investigating a better cooperation strategy between the slave processes, in particular in the context of graphics processing unit (GPU) computing.

Acknowledgements. This work is partly supported by the ANR (French Research National Agency) funded project FiCOLOFO ANR-10-BLA-0214. We gratefully acknowledge High Performance Computing Unit (UCI) of University of Oran for making available for us computing facilities.

References

1. Arnborg, S., Corneil, D.G., Proskurowski, A.: Complexity of finding embeddings in a k-tree. SIAM Journal on Algebraic and Discrete Methods 8, 277–284 (1987)
2. Bensana, E., Lemaître, M., Verfaillie, G.: Earth observation satellite management. Constraints 4(3), 293–299 (1999)
3. Cabon, B., de Givry, S., Lobjois, L., Schiex, T., Warners, J.P.: Radio link frequency assignment. Constraints 4(1), 79–89 (1999)
4. Crainic, T.G., Gendreau, M., Hansen, P., Mladenovic, N.: Cooperative parallel variable neighborhood search for the p-median. Journal of Heuristics 10(3), 293–314 (2004)
5. de Givry, S., Schiex, T., Verfaillie, G.: Exploiting tree decomposition and soft local consistency in weighted csp. In: AAAI, pp. 22–27. AAAI Press (2006)
6. Dechter, R., Pearl, J.: Tree clustering for constraint networks. Artificial Intelligence 38(3), 353–366 (1989)
7. Carlson, C.S., et al.: Selecting a maximally informative set of single-nucleotide polymorphisms for association analyses using linkage disequilibrium. American Journal of Human Genetics 74(1), 106–120 (2004)
8. Fontaine, M., Loudni, S., Boizumault, P.: Exploiting tree decomposition for guiding neighborhoods exploration for VNS. RAIRO Operations Research 47(2), 91–123 (2013)
9. Gottlob, G., Lee, S.T., Valiant, G.: Size and treewidth bounds for conjunctive queries. In: Paredaens, J., Su, J. (eds.) PODS, pp. 45–54. ACM (2009)
10. Harvey, W.D., Ginsberg, M.L.: Limited discrepancy search. In: IJCAI, pp. 607–615. Morgan Kaufmann (1995)
11. Larrosa, J., Schiex, T.: In the quest of the best form of local consistency for Weighted CSP. In: IJCAI, pp. 239–244 (2003)
12. Loudni, S., Boizumault, P.: Combining VNS with constraint programming for solving anytime optimization problems. European Journal of Operational Research 191, 705–735 (2008)
13. Marinescu, R., Dechter, R.: AND/OR branch-and-bound search for combinatorial optimization in graphical models. Artificial Intelligence 173(16-17), 1457–1491 (2009)
14. Mladenovic, N., Hansen, P.: Variable neighborhood search. Computers and Operations Research 24, 1097–1100 (1997)
15. Pearl, J.: Probabilistic inference in intelligent systems. In: Networks of Plausible Inference. Morgan Kaufmann (1998)
16. Rish, I., Dechter, R.: Resolution versus search: Two strategies for SAT. Journal of Automated Reasoning 24(1/2), 225–275 (2000)
17. Robertson, N., Seymour, P.D.: Graph minors. ii. algorithmic aspects of tree-width. Journal of Algorithms 7(3), 309–322 (1986)
18. Sánchez, M., Allouche, D., de Givry, S., Schiex, T.: Russian doll search with tree decomposition. In: Boutilier, C. (ed.) IJCAI, pp. 603–608 (2009)
19. Sánchez, M., de Givry, S., Schiex, T.: Mendelian error detection in complex pedigrees using weighted constraint satisfaction techniques. Constraints 13(1-2), 130–154 (2008)
20. Tarjan, R.E., Yannakakis, M.: Simple linear-time algorithms to test chordality of graphs, test acyclicity of hypergraphs, and selectively reduce acyclic hypergraphs. SIAM Journal on Computing 13(3), 566–579 (1984)

GeNePi: A Multi-Objective Machine Reassignment Algorithm for Data Centres

Takfarinas Saber[1], Anthony Ventresque[1],
Xavier Gandibleux[2], and Liam Murphy[1]

[1] Lero@UCD, School of Computer Science and Informatics,
University College Dublin, Ireland
takfarinas.saber@ucdconnect.ie, {anthony.ventresque,liam.murphy}@ucd.ie
[2] Faculty of Science, University of Nantes, France
xavier.gandibleux@univ-nantes.fr

Abstract. Data centres are facilities with large amount of machines
(i.e., servers) and hosted processes (e.g., virtual machines). Managers of
data centres (e.g., operators, capital allocators, CRM) constantly try to
optimise them, reassigning 'better' machines to processes. These managers usually see better/good placements as a combination of distinct
objectives, hence why in this paper we define the data centre optimisation problem as a *multi-objective machine reassignment problem*. While
classical solutions to address this either do not find many solutions (e.g.,
GRASP), do not cover well the search space (e.g., PLS), or even cannot operate properly (e.g., NSGA-II lacks a good initial population), we
propose *GeNePi*, a novel hybrid algorithm. We show that GeNePi outperforms all the other algorithms in terms of quantity of solutions (nearly
6 times more solutions on average than the second best algorithm) and
quality (hypervolume of the Pareto frontier is 106% better on average).

Keywords: Data Centres, Machine Reassignment, Evolutionary Algorithms, Multi-Objective Optimisation.

1 Introduction

Data centres are facilities dedicated to hosting many computer resources, and
while they have been around for decades, they are now the centre of attention
as they are increasingly the crucial element of our digital lives (e.g., for the
Cloud). These data centres evolve constantly as for instance machine age and are
eventually decommissioned, new ones (more powerful) are bought regularly, and
processes hosted are updated to potentially more greedy ones. Managers of data
centres have to adapt their systems to these evolutions and migrate processes
from one machine to another according to technical and non-technical reasons,
what we call *reassignment of processes to machines*. For instance, managers may
want to increase the reliability of their data centre and move workload from
overloaded machines to less loaded and/or more powerful ones. Often, they also
try to move services to power efficient machines, in order to lower the cost and
environmental impact of the data centres.

M.J. Blesa, C. Blum, and S. Voß (Eds.): HM 2014, LNCS 8457, pp. 115–129, 2014.
© Springer International Publishing Switzerland 2014

One problem is that machines can range to up to tens of thousands (e.g., OVH, a European leader in the domain, have 150,000 servers in 12 data centres[1]), and services up to millions (e.g., VMware ESX accepts up to 320 VMs per host). At this scale, any instance of the reassignment problem becomes a challenge to the existing heuristics and solvers, and finding the 'best' (re-)assignment an illusion. Another problem is that, as we mentioned in the previous paragraph, managers have different perspectives on what is a 'good' solution, and ranking all the solutions according to a single utility function (e.g., minimising energy consumption) is probably not relevant.

This is a perfect example of a problem where *multi-objective decision making* makes sense: an optimisation problem with various independent objectives that only decision makers can compare – possibly collectively. For instance, Xi et al. [1] describe such an enterprise environment where managers of virtual data centres have various perspectives when it comes to placement decisions. Hence we call the problem we address in this paper *multi-objective optimisation for the machine reassignment problem*. While this problem has been addressed in the context of machine assignment [2], or for dynamic assignment of small amount of machines [3], it has not been in itself the topic of research in the past. In this paper we identify three objectives for the problem: (i) reliability, i.e., a penalty is given to assignments that load too much the machines; (ii) migration, i.e., assignments that move processes too much (especially to remote locations) are penalised; and (iii) electricity: trying to obtain assignments that minimise the (electrical) cost of running the data centre.

In this paper we show that the classical solutions do not perform well against this problem, in terms of the number of non-dominated solutions found (the *quantity* of solutions) or the hypervolume [4] of the search space area defined by the Pareto frontier (the *quality* of the solutions). Pareto Local Search (PLS) [5,6,7] usually finds solutions but they are grouped in one area of the search space (small hypervolume) and it is a slow algorithm – these are the expected behaviour of this algorithm. NSGA-II [8] needs a good initial population in order to operate properly, while here it gets only one solution: the initial assignment. GRASP [9] does not perform well in such large search spaces and ends up trying a lot of non-feasible settings, eventually finding few or no solutions. We then propose a novel hybrid algorithm called *GeNePi*, using successfully three steps: a first step (inspired from GRASP) to explore quickly all the search space, a second (using NSGA-II) to introduce some variety and quality in the solutions and a last one (PLS-based) to increase the number of solutions. GeNePi outperforms all the state-of-the-art other algorithms (the previously mentioned ones and some classical bin packing ones), finding nearly 6 times more non-dominated solutions on average and covering the search space better (106% better on average).

In the rest of this paper we first give a problem definition, with the constraints and the three objectives that we identified as the most relevant (Section 2). Then we describe GeNePi, our algorithm for solving this multi-objective machine reassignment problem (Section 3). After this, section 4) proposes an evaluation

---

[1] Source: http://www.ovh.com/fr/backstage/ – accessed on 23/11/2013.

of GeNePi against the state-of-the-art other algorithms. Finally we make some concluding remarks (Section 5).

2   Problem Definition

The *Multi-Objective Machine Reassignment Problem* consists in optimising the usage of a set of machine $\mathcal{M}$ according to various objectives. More specifically, reassignment in general seeks to find a new machine $M(p)$ for every process p in the system, initially placed in machine $M_0(p)$, satisfying the constraints of the system while its multi-objective version tries to find non-dominated (better than every other solution in some directions of the space). In some cases $M_0(p) = M(p)$, which means that the process p does not move during the reassignment.

The machine reassignment problem can be seen as a complex (i.e., with more constraints such as dissemination and dependency) instance of *d-dimensional vector bin packing* [10,11] with each machine a *d-dimensional bin*, such that d is the number of the resources. The aim is to place the processes in these machines, such a way they satisfy their capacities and they minimise the number of bins used. This problem is $NP - Hard$ and it has drawn lot of attention [12]; in this work we do not consider the scheduling aspects of it (*bin repacking scheduling problem* [13]). The model we describe below is loosely inspired by several work (e.g., [14] for a linear model), among which the problem definition of the ROADEF challenge [15] has an important place.

2.1   Reassignment Problem

A machine $m \in \mathcal{M}$ belongs to a location $l \in \mathcal{L}$ (the site where the server is located). It is also in a neighbourhood $N(m) \subseteq \mathcal{M}$, which represents a set of machines with which it is linked to by fast connections or with which it shares the same protocol. Each machine belongs to one and only one location and one neighbourhood. Every m has also several resources $r \in \mathcal{R}$ (e.g., RAM, CPU, disk), in limited capacities $Q_{m,r}$. We consider that the quantity of resource r that the process p needs is fixed to $d_{p,r}$ and corresponds to a VM parameter/SLA[2]. The first constraint regards the number of processes a machine can host:

$$\sum_{p \in \mathcal{P} \mid M(p)=m} d_{p,r} \leq Q_{m,r}, \quad \forall m \in \mathcal{M}, \forall r \in \mathcal{R} \tag{1}$$

Some resources are called *transient*: $r \in \mathcal{TR} \subseteq mathcalR$. Such resources (e.g., RAM and disk) are needed on both machines during a new assignment, as the processes use the resources on both machines during the migration.

$$\sum_{p \in \mathcal{P} \mid M_0(p)=m \vee M(p)=m} d_{p,r} \leq Q_{m,r}, \quad \forall m \in \mathcal{M}, \forall r \in \mathcal{TR} \tag{2}$$

---

[2] a Service-Level Agreement (SLA) is a contract agreed between a data centre provider and a customer which describes the service provided (e.g., allocated resources, time to recover after an outage).

Other resources are called *non-transient*: $r \in \mathcal{NR} = \mathcal{R} \setminus \mathcal{TR}$.

Services/applications are often multi-tier (e.g., to separate concerns) and replicated (for performance and security reasons), so it is realistic to assume here that processes (the atomic element of workload) are organised by services. It is a common for services to have an *anti-cohabitation* constraint [16], i.e., the processes composing a service cannot share the same host – for some reliability, security and performance reasons. Let $\mathcal{P}$ be the set of processes and $\mathcal{S}$ the set of services, then the *anti-cohabitation* constraint can be expressed as in (3).

$$\forall p_i, p_j \in \mathcal{P}, i \neq j, \forall s \in \mathcal{S}, (p_i, p_j) \in s^2 \Rightarrow M(p_i) \neq M(p_j) \tag{3}$$

For the same reasons of reliability, security and performance, services require that the number of locations hosting at least one process has to be greater than a certain number, caller *spread number*. This allows increasing the resilience in case of failure of a data centre: the bigger the spread number, the safer the service.

$$\sum_{l \in \mathcal{L}} \min \left(1, |\{p \mid p \in s \wedge M(p) \in l\}|\right) \geq spreadNumber_s, \quad \forall s \in \mathcal{S} \tag{4}$$

Services can also depend on each other and in this case the processes of these services need to be close to each other – to increase the performance of the system. We note this dependency of services $\hookrightarrow$. Of course, as the dependencies between services can be complex, the assignment can be tricky: a process $p \in \mathcal{P}$, belonging to service $s_i \in \mathcal{S}$ which is dependent on service $s_j \in \mathcal{S}$ and service $s_k \in \mathcal{S}$, needs to be assigned to a machine in $N(m)$ with $\exists p' \in s_j \cap s_k m = M(p')$.

$$\forall s_i, s_j \in \mathcal{S}, \ s_i \hookrightarrow s_j \implies \forall p_u \in s_i, \ \exists p_v \in s_j \mid N(M(p_u)) = N(M(p_v)) \tag{5}$$

Figure 1 shows graphically a scenario (i.e., instance and initial solution) of the problem. Note that resource capacities and demands are not represented here to make it simpler to understand.

Definition 1 (Machine Reassignment). *An assignment A of processes to machines is a mapping: $A : \mathcal{P} \mapsto \mathcal{M}$, such that $A(p, \mathcal{M}) \to m$, which satisfies all the previous constraints 1, 2, 3, 4 and 5.*

A reassignment is a function that modifies an initial assignment: $ReA : A \mapsto A$ and gives a new assignment of processes to machines.

2.2   Objectives

As said in the introduction, there are several perspectives on the best optimisation, which translate in our case into several objectives. Some studies [17] show that a large number of objectives decreases drastically the performance of evolutionary algorithms, and that decision makers tend to favour small number of dimensions. We have then decided to focus only three objectives that seem to make the most sense from the literature and discussions with industrials: reliability, migration and electricity costs.

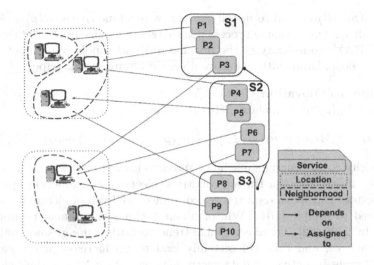

Fig. 1. Simple scenario of a correct assignment of processes to machines (spread= 2)

There are many elements that can help data centre operators to predict the risk of failure of a server: to name a few the age of a machine, the vendors of its parts (e.g., processor maker) and the past history of similar machines. They are complex to collect and understand, and we do not know exactly how to process them to obtain an objective that the data centre operators and decision makers could use (the literature seems uncertain on the matter [18]). One thing we know is that as opposed to the risk of failure, the *reliability* is easier to compute and gathers less questions.Machines do operate better when they are not too loaded, and reliability can be estimated through the load: the more loaded a machine, the greater the risk of performance issues or failures.

Definition 2 (Reliability Cost). *A machine $m \in \mathcal{M}$ is reliable if it is not loaded more than a reliability value $\rho(m, r)$ for each resource $r \in \mathcal{R}$, and we compute a reliability cost of m, $\rho(m)$, as:*

$$\rho(m) = \sum_{r \in \mathcal{R}} \max \left(0, \sum_{p \in \mathcal{P}|M_o(p)=m \vee M(p)=m} d_{p,r} - \rho(m,r) \right) \quad (6)$$

If the *safety capacity* of m for the resource r is higher than the sum of the demands, then it does not impact the safety of the machine. Note that this definition is inspired by the concept of *safety capacity* introduced in [19]: if one or several resources of a machine are over-loaded then the machine may not be able to satisfy its SLAs.

Migrating a process has a cost which is often neglected by research in the area but is well known by practitioners [20]. Basically, this consists in the time needed to prepare a process p for a migration ($\mu_1(p, M_o(p))$), to transfer p

$(\mu_2(p, M_o(p), M(p)))$ and to install p on a new machine $(\mu_3(p, M(p)))$. All these costs are dependent on some process parameters (e.g., size of the data stored on disk and RAM, complexity of the installation) and topology parameters (e.g., number of hops, bandwidth), that we do not evaluate in this paper.

Definition 3 (Migration Cost). *The cost of migrating a process $p \in \mathcal{P}$ from a machine $M_0(p)$ to a machine $M(p)$ is defined:*

$$\mu(p, M_o(p), M(p)) = \mu_1(p, M_o(p)) + \mu_2(p, M_o(p), M(p)) + \mu_3(p, M(p)) \qquad (7)$$

Electricity cost of running machines accounts for up to 50% of their operating costs [21] and it is a burden for countries' electricity production systems: in 2007, Western European data centres consumed 56 TWh of electricity, and this is expected to double (104 TWh, or about 4 times the annual production or Ireland) by 2020 [22]. There is a global trend towards more greener and power-aware practices, and this will certainly lead to an increase in the electricity price and other incentive for data centre managers to minimise their electricity consumption. Modelling electricity cost is complex but we follow the general assumption that states that it is a linear function of its CPU usage [23,24]. We then just define two parameters, α_m (linear factor) and β_m (fixed cost of running m with n load on the CPU) for every machine m. This does not take into account other elements that may be relevant but are somehow out of the scope of our study here (e.g., cooling of data centres).

Definition 4 (Electricity Cost). *The electricity cost of a machine $m \in \mathcal{M}$ in the location $l \in \mathcal{L}$ depends on the variables α_m, β_m (electricity consumption constants) and γ_l (electricity cost in l), and is expressed by the following formula:*

$$\epsilon(m) = \begin{cases} \gamma_l \times \left(\alpha_m \times \displaystyle\sum_{p \in \mathcal{P} \mid M(p)=m} d_{p,CPU} + \beta_m \right) & \text{if } m \text{ is running} \\ 0 & \text{otherwise} \end{cases} \qquad (8)$$

3   Description of Our Solution: GeNePi

GeNePi applies successively three (modified) optimisation algorithms: GRASP, NSGA-II and PLS. This idea of using three steps for approximate resolution [25] is new in the domain of data centres optimisation.

3.1   Ge: A Variant of the Constructive Phase of GRASP

We use a variant of the constructive phase of *Greedy Randomized Adaptive Search Procedure* [9] (GRASP). Solutions are generated by trying to reassign processes one after the other, according to a greedy heuristic which is slightly relaxed to include a random factor. This method is commonly used for combinatorial problems, and applied to get some quick initial solutions with good objectives. After ranking the processes according to their dependencies and their needs of resources, they are selected one by one. A decision of reassigning one per cent of the processes from their initial hosts has been taken, because of the tightness of transient resource constraints that limits the number of reassignments. The choice of the reassignment of every process is based on a linear combination of the three utility/objective functions (one per objective). Even if a linear combination of these utility functions allows us to go beyond the objective types barrier, its static definition induces getting solutions with a same objectives level of interest. This behaviour goes against the aim of a multi-objective optimisation. That is why we adopted a panel of triplet weights $(\lambda_i, \lambda_j, \lambda_k)$ in $]0,1]^3$, with $\lambda_k = 1 - \lambda_i - \lambda_j$. They are chosen in such a way they cover a maximum search space by optimising the objectives separately in addition to their trade-offs. They will be used to introduce a diversification in the interest of each objective, ensuring a trade-off between them. The random part of GRASP lays in the assignment of a machine to each process, at each iteration. For each process, a set of assignable machines that respect the constraints is computed, and a value of interest is given to each machine by a weighted sum: (U_i): $\sum_{i=1}^{3} \lambda_i U_i$, which creates a set of machine with a utility lower than or equal to $(minUtility + (1 - r) * [maxUtility - minUtility])$, with $r \in [0,1]$. A random machine is selected from this eligible set to assign the process to it. During the assignment, it may happen that a process has no machine able to host it. The solution is declared infeasible, and removed from the initial solutions. Globally, at the end of this step, we expect to have a set of decent solutions spread over the search space.

3.2   Ne: NSGA-II

We use for this step a genetic algorithm called *Non-dominated Sorting Genetic Algorithm-II* [8] (NSGA-II). This step is useful for the improvement of the *Pareto set*[3] obtained from the first step. This metaheuristic allows to get a good dissemination of the solutions around the Pareto frontier and prevent their accumulation in some area of the search space. Hopefully, it allows GeNePi getting a smooth frontier and increases the number and the quality of the non-dominated solutions. It is a genetic algorithm, i.e., it runs an evolutionary process which matches individuals (i.e., solutions or assignments) at each generation and mixes their features (as the biological evolution would do with genes). The two main actions are *crossover* which mixes genes from two parents, and *mutation* that creates randomly individuals with new features. There exists several ways of doing crossovers, which is more or less a cut and paste operation where assignments in the set of actual solutions are split into regular length segments and swapped with one another [26]. In our case crossovers consider the exchange of services (group of processes) rather than blocks of process assignments – that minimise the number of bad crossovers. Of course the diversity is less than with crossovers on processes, but we compensate with a bigger probability of mutations (i.e., random assignments in solutions to see whether this improve the utilities). After a

---

[3] Pareto set: a set of non-dominated solutions (i.e., better than all other solutions in one or more objectives).

generation has "passed", some new individuals are kept (usually the fittest, those with the best objective values: low domination rank, but also some other that allow to introduce some variety: high crowding-measure [8]), and others are suppressed. Hence the global population of assignments only improves (descendants worse than their parents are likely to be suppressed). Beside, last generations tend to be well distributed over the Pareto frontier.

3.3   Pi: A Pareto Local Search

Finally, we try to improve the Pareto set by using a *Pareto Local Search* [5,6,7] (PLS). It consists in applying several local search operators on the solutions belonging to the *Pareto* frontier. Few simple moves are chosen to analyse the neighbourhood of actual solutions: (i) swap, i.e., taking two processes and exchanging their assignment; (ii) 1-exchange, where one process at a time is selected and reassigned to any machine that accepts it; (iii) shift, where processes belonging to the same service exchange their assignments (which maintains the satisfaction on the dependency constraints). These moves allow probing of a large neighbourhood around the current solutions, which may generate some redundancies if the solutions are close of one another. To overcome this problem, we generate boxes by clustering solutions, and apply a local search to the most isolated solution in each of them (i.e., has the largest crowding-measure value). Only one neighbourhood is generated for every selected solution at every iteration, even if new interesting solutions have been found. This balances the improvement and reduces the execution time as redundancy is less likely.

4   Evaluation

In this section, we evaluate the performance of our solution against other state-of-the-art multi-objective and reassignment solutions, using several metrics: time, quantity (number of solutions) and quality of solutions (hypervolume). We create a benchmark inspired by the ROADEF Challenge 2012 [19].

4.1   Experimental Setups

The ROADEF challenge 2012 is particularly suited to our needs, as it is rather realistic (proposed by Google, who claim it represents accurately some of their data centres) and it is quite comprehensive: lot of resources for the machines/processes while many papers in the area only consider two (namely, RAM and CPU), reasonably high number of machines and processes, complex dependencies and constraints on the services and processes which make the assignments not straightforward. The ROADEF dataset distinguishes three categories of instances (a_1 are considered 'easy', a_2 'medium' and b 'hard').

In this paper, we pick up 14 instances (see Figure 1), leaving only the biggest ones. We have added variables α_m and β_m to each machine $m \in \mathcal{M}$, and γ_l for every location $l \in \mathcal{L}$ in order to include electricity consumption. All tests were made on a 4 cores Intel Xeon 3.10GHz CPU, with 8GB of RAM, running Ubuntu 12.4 LTS 64-bits.

Table 1. The dataset used for our evaluation (ID and size of the different instances) and execution time of GeNePi on them

Instance	# Resources	# Machines	# Services	# Processes	Execution Time (s)
a_1_1	2	4	79	100	1.58
a_1_2	4	100	980	1,000	3,108
a_1_3	3	100	216	1,000	441
a_1_4	3	50	142	1,000	309
a_1_5	4	12	981	1,000	332
a_2_1	3	100	1,000	1,000	3,905
a_2_2	12	100	170	1,000	600
a_2_3	12	100	129	1,000	695
a_2_4	12	50	180	1,000	342
a_2_5	12	50	153	1,000	347
b_1	12	100	2,512	5,000	14,990
b_2	12	100	2,462	5,000	10,028
b_3	6	100	15,025	20,000	39,595
b_4	6	500	1,732	20,000	63,534

4.2   Metrics

Comparing multi-objective optimisation approaches is complex as the set of solutions they give on a problem can be seen from different perspectives: coverage, closeness to the Pareto frontier, variety, and many more [27]. The problem probably roots in the fact that the Pareto frontier is unknown most of the time, and that the different objectives cannot be taken in isolation to give the quality of any solution. In this paper, we made the decision to take only few unary operators as metrics (see other studies for a more comprehensive study of the various possible operators [28]): unary as they take a set of solutions and give a single value, which allows comparing the different approaches.

The first metric we use is the number of *non-dominated (efficient) solutions* and we refer to it as the quantity of solutions. Finding a large number of solutions is always better as it provides more alternatives to the decision makers.

The other metric is the *hypervolume* [4] (also known as the S metric). We sometimes call it quality of the solutions. This is a widely used metric in the area of optimisation to evaluate the performance of multi-objective algorithms that aims at understanding how the output sets are spread in the different dimensions. In short, the hypervolume measures the space (in the n dimensions of the n objectives) defined by the set of non-dominated solutions and a reference point, picked in the space as far as possible from the Pareto frontier. The bigger the hypervolume, the more interesting are the solutions in the found non-dominated solution set, as they increase the dominated area. In formal terms, this is proven by Fleischer [29] who states that the maximisation of the hypervolume is equivalent to finding the optimal Pareto frontier. Note that in order to compare the result sets of different algorithms, we use the same reference points for each instance of the multi-objective machine reassignment problem. A last remark is that although hypervolume is obviously impacted by the random seeds applied to algorithms, a preliminary study we conducted did not give us any sense that the relative results of the different algorithms would vary. In other words, good algorithms are always good, bad are always bad, whatever the seed - in particular GeNePi always gives the best hypervolume values. We hence use the same seed for each algorithms, but plan to study in more details the impact of each technique on the Pareto front in some future work.

4.3   Algorithms

We compare our solution to three different types of algorithms, running for the same period of time. The first algorithms are from the *First Fit family*. These heuristics are designed for Vector Bin Packing [30] and they are considered efficient. We chose among them the First Fit (FF) which selects the first machine that fits for every process in a sequence; the Random Fit (RF) which selects randomly a machine among those which fit; and the First Fit Descent Bin-Balancing (BB) which selects the least loaded machine for each process.

The second set of algorithms is the state-of-the-art solutions from the multi-objective optimisation field. The first of them is GRASP in its original definition, i.e., the choice of reassigning is based on a uniform probability distribution on the eligible machines. We also run the first step of GeNePi (Ge) as it is a variation on GRASP that we expect is better than GRASP for our scenario. We also try NSGA-II where we reserve a third of the execution time to GRASP in order to create an initial population and run NSGA-II in the two remaining thirds of the execution time. The last Algorithm is a Pareto Local Search (PLS), with a number of boxes at every iteration equal to the number of solutions in the non-dominated solution set.

4.4   Tuning the Steps of GeNePi

Each of the three steps composing GeNePi has several parameters that need to be tuned, and globally we need to decide how many iterations or how much time we allocate to each of them to make the best use of each. The tuning has been performed on the instance a_1_5.

The first step of GeNePi is Ge (based on GRASP), which has only one value to tune: α, the factor leading to more randomised greedy search (bigger α) or local search (smaller α). We conducted a thorough evaluation of the impact of different values of α from 0.05 to 0.95 (repeated 10 times). The best value of α seems to be 0.6 regardless to the number of iterations. For Ne (i.e., NSGA-II), we combined 9 possible values $\{0.1, 0.2, \ldots, 0.9\}$ for P_c and P_m, obtaining 81 different variations of the parameters (we again run 10 times each combination). We realise that P_c values between 0.6 and 0.8 give better results, while the impact of P_m seems less important (values of P_m between 0.1 and 0.3 giving slightly better results though). We then decided to use $P_c = 0.6$ and $P_m = 0.2$. Pi has only one parameter that we can tune here: the number of zones (boxes) that it can explore. This number of zones has an impact on the quality of the Pareto frontier, and hence on the hypervolume. A small number of zones means less neighbourhood probing, but also less redundancy and execution time, while more zones allow to analyse more neighbourhoods (and to find more solutions) but there is a cost in redundancy and execution time. 10 seems to us a good trade-off between probing a large search area and reducing the execution time. Table 2 summaries the tuned parameters for each part of GeNePi.

GeNePi aims at providing decision makers with an important set of good solutions, covering the solutions space, and in a reasonable time. These two ideas (quality and time) seem to be incompatible, but they just force to consider time in a different way. In particular, Ne/NSGA-II needs to have a set of good initial solutions, and then we have to make sure Ge/GRASP has enough time. It appeared from tests that served to define α that a number of iterations from 100 to 500 lead to practically the same hypervolume. This is why; we picked 100 as the number of iteration for Ge. The good number of iterations for Ne/NSGA-II is much trickier to find as it depends greatly on the quality of the initial population. It seems, experimentally, that 100 iterations with a population size of 50 give good results, so this is what we use for Ne/NSGA-II. Pi/PLS is the most time consuming part, we use it only for one iteration in order to get a smooth Pareto.

Table 2. Parameters for the different steps of GeNePi after a tuning study

Ge (1st step - GRASP)		Ne (2nd step - NSGA-II)		Pi (3rd step - PLS)	
α	0.6	Probability of crossover	0.7	# zones (# boxes)	10
$|\Lambda|$	4	Probability of mutation	0.3	# iterations	1
# iterations	100	Size of population	50		
		# iterations	100		

4.5 Results

Table 3 shows that GeNePi outperforms notably all the other algorithms, both in terms of number of solutions (5.84 times more solutions found than the second best on average) and hypervolume (106% better on average). We notice that solutions from the first fit family have acceptable results only for some instances (a_1_1, a_1_5 and a_2_1). In general those algorithms favour the reassignment of the processes, with the side effect that more transient resources are consumed and more anti-cohabitation and dependency constraints are violated. That is why they perform better with instances that have larger transient resource capacities and a ratio between the number of processes and services close to one – this corresponds exactly to the three instances mentioned above. The same behaviour is observed for GRASP, which tends to reassign processes instead of keeping them on their initial assignment, as a decision is made based on a basic draw among several relevant machines for every process, based on a utility function. Hence the probability of choosing the initial assignment is low and GRASP tries a lot of solutions that end up being infeasible. NSGA-II, being dependent on the quality of the initial population, performs badly – although we give a

Table 3. Summary of solutions found and hypervolume (in brackets) for the various algorithms and the various instances. For both metrics, the higher the better. We put in bold the best values for each instance.

	RF	FF	BB	GRASP	Ge	NSGA	PLS	GeNePi
a_1_1	4(2.6)	10(2.95)	87(3.73)	20(2.73)	42(3.6)	14(2.8)	10(2.4)	**193(3.95)**
a_1_2	1(7.49)	1(7.49)	1(7.49)	1(7.49)	26(8.47)	1(7.49)	2(7.49)	**133(9.14)**
a_1_3	1(4.17)	1(4.17)	1(4.17)	1(4.17)	19(4.27)	1(4.17)	2(4.17)	**62(4.32)**
a_1_4	1(9.72)	1(9.72)	1(9.72)	1(9.72)	40(11.1)	1(9.72)	2(9.72)	**187(12.2)**
a_1_5	4(2.51)	10(2.51)	2(2.45)	14(2.59)	49(2.74)	16(2.57)	32(2.52)	**66(2.78)**
a_2_1	33(4.86)	41(4.95)	1(4.57)	69(5.41)	57(5.43)	71(5.46)	4(4.61)	**227(5.68)**
a_2_2	1(1.33)	1(1.33)	1(1.33)	1(1.33)	22(1.55)	1(1.33)	2(1.33)	**171(1.76)**
a_2_3	1(2.02)	1(2.02)	1(2.02)	1(2.02)	30(2.36)	1(2.02)	67(2.04)	**250(2.67)**
a_2_4	1(6.42)	1(6.42)	1(6.42)	1(6.42)	28(7.62)	1(6.42)	2(6.42)	**374(8.63)**
a_2_5	1(9.91)	1(9.91)	1(9.91)	1(9.91)	28(10.3)	1(9.91)	2(9.91)	**245(11)**
b_1	1(8.2)	1(8.2)	1(8.2)	1(8.2)	27(8.34)	1(8.2)	39(8.34)	**207(8.49)**
b_2	1(1.43)	1(1.43)	1(1.43)	1(1.43)	23(1.48)	1(1.43)	2(1.43)	**300(1.53)**
b_3	1(6.25)	1(6.25)	1(6.25)	1(6.25)	20(6.27)	1(6.25)	108(6.27)	**162(6.3)**
b_4	1(3.65)	1(3.65)	1(3.65)	1(3.65)	22(3.67)	1(3.65)	3(3.67)	**118(3.7)**

partial result of GRASP to help it at the start. This is a major (but well known) drawback for this algorithm, especially for our scenario for which NSGA-II is clearly not fitted. The results for PLS are contrasting as they can be good in terms of quantity (better than Ge at times) but are poor in terms of quality – hypervolume values for PLS are always among the lowest. This comes from the fact that PLS searches for possible solutions locally, and may find some, but they are similar to the original ones and do not increase the diversity of the solutions set. For a multi-objective problem like ours, PLS is then not fitted either. Ge, the first step of GeNePi gets a good hypervolume but not an outstanding number of solutions. This was expected as it is only an improvement of GRASP which itself suffers from a lack of solutions. GeNePi is by far the best algorithm, and we explain it by the composition of elements: Ge (i.e., modified GRASP) finds a large number of solutions, allowing NSGA-II (the second step to operate properly and finding new solutions that compromise all the objectives, while PLS, the last step, increase the number of solutions around the previously found ones.

Table 1 shows that GeNePi works in a short time for the easy and medium instances, and in a reasonable time (for our scenario) for the bigger ones. The 17 hours of running GeNePi for the biggest instance we consider (b_4) are totally justified if this can save money, increase the reliability and do not put the data centre at risk by performing too many migrations. Especially as GeNePi can give 118 solutions for this instance, i.e., 118 options for the operators to make the most informed decision. To give the reader a sense of what happens during GeNePi and specially the impact of the three phases, we plot the improvement curve of the instance a_1_5 (see Figure 2). Each point corresponds to one or several new non-dominated solutions found (with the timestamp of this new solution in x-axis and the new hypervolume of the solution set in y-axis). We can see that GRASP finds solutions quickly (9 s.) and the rise in the hypervolume signifies

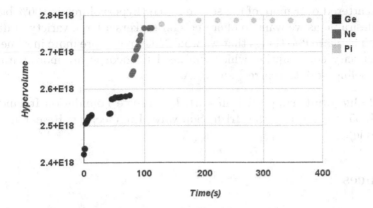

Fig. 2. Improvement curve of GeNePi on the instance a_1_5. Each point is a new solution (or a set of new solutions).

that they are distributed in the search space (which is good). GRASP continues after the first 9 s. and finds new solutions, but we notice a quasi-stagnation of the hypervolume after 46 s. NSGA-II finds group of solutions, each group improving significantly the hypervolume: they correspond to a new search areas in the space discovered by the first solution of the group and improved by the others. PLS takes more time than the other steps, but it finds some new solutions and bring a little improvement in terms of hypervolume.

5 Conclusion

Reassigning processes to servers automatically is complex (lot of dimensions and constraints), large scale for most of the real instances (data centres are usually big computing facilities) and needs to consider different objectives. In this paper we define the multi-objective machine reassignment problem and propose an hybrid solution using successively three optimisation steps: GeNePi. Multi-objective approaches are good when the set of possible solutions is large and extracting the 'best solution' is difficult. In this case, the system needs to be assisted by decision makers who can evaluate the different solutions with respect to their value in the different dimensions of the problem. Here, we defined the machine reassignment in the three dimensional space defined by: (i) reliability of the assignment, (ii) migration cost of the assignment, and (iii) energy consumption. Our solution, GeNePi, is based on three optimisations algorithms: Ge, a variant of the constructive phase of GRASP, which aims at finding an initial population with solutions representing every objective; Ne, based on a genetic algorithm called NSGA-II that mixes solutions of the initial population and tries to find new solutions (and more diverse ones); and Pi a local search that looks for more solutions in the neighbourhood of those that GeNePi has already found. We showed in a large experimental validation that GeNePi outperforms other state-of-the-art solutions: it finds 5.84 times more good (non-dominated) solutions that are scattered over more of the search space (hypervolume is 106% better) – which is desirable as we want to offer decision makers a large variety of different solutions. There are two things that we would like to explore more in some future work: electricity consumption which will need to incorporate more parameters (such as cooling of data centres) and SLAs.

Acknowledgement. supported, in part, by Science Foundation Ireland grant 10/CE/I1855 to Lero - the Irish Software Engineering Research Centre (www.lero.ie).

References

1. Li, X., Ventresque, A., Stokes, N., Thorburn, J., Murphy, J.: ivmp: an interactive vm placement algorithm for agile capital allocation. In: CLOUD, pp. 950–951 (2013)
2. Mills, K., Filliben, J., Dabrowski, C.: Comparing vm-placement algorithms for on-demand clouds. In: CloudCom, pp. 91–98 (2011)
3. Xu, J., Fortes, J.: A multi-objective approach to virtual machine management in datacenters. In: CAC, pp. 225–234 (2011)
4. Zitzler, E., Thiele, L.: Multiobjective optimization using evolutionary algorithms - a comparative case study. In: Eiben, A.E., Bäck, T., Schoenauer, M., Schwefel, H.-P. (eds.) PPSN 1998. LNCS, vol. 1498, pp. 292–301. Springer, Heidelberg (1998)
5. Angel, E., Bampis, E., Gourves, L.: A dynasearch neighborhood for the bicriteria traveling salesman problem. In: Metaheuristics for Multiobjective Optimisation, pp. 153–176 (2004)
6. Basseur, M.: Design of cooperative algorithms for multi-objective optimization: application to the flow-shop scheduling problem. In: 4OR, pp. 255–258 (2006)
7. Alsheddy, A., Tsang, E.E.: Guided pareto local search based frameworks for biobjective optimization. In: CEC (2010)
8. Deb, K., Pratap, A., Agarwal, S., Meyarivan, T.: A fast and elitist multiobjective genetic algorithm: Nsga-ii. In: TEVC, pp. 182–197 (2002)
9. Feo, T.A., Resende, M.G.: Greedy randomized adaptive search procedures. In: JGO, pp. 109–133 (1995)
10. Gabay, M., Zaourar, S.: A GRASP approach for the machine reassignment problem. In: EURO (2012)
11. Bansal, N., Caprara, A., Sviridenko, M.: Improved approximation algorithms for multidimensional bin packing problems. In: FOCS, pp. 697–708 (2006)
12. Batu, T., Rubinfeld, R., White, P.: Fast approximate PCPs for multidimensional bin-packing problems. In: Information and Computation, pp. 42–56 (2005)
13. Hermenier, F., Demassey, S., Lorca, X.: Bin repacking scheduling in virtualized datacenters. In: Lee, J. (ed.) CP 2011. LNCS, vol. 6876, pp. 27–41. Springer, Heidelberg (2011)
14. Mehta, D., O'Sullivan, B., Simonis, H.: Comparing solution methods for the machine reassignment problem. In: Lee, J. (ed.) CP 2011. LNCS, vol. 6876, pp. 782–797. Springer, Heidelberg (2011)
15. Google/roadef/euro challenge 2012: Definition of the machine reassignment problem (2012),
http://challenge.roadef.org/2012/files/problem_definition_v1.pdf

16. Bin, E., Biran, O., Boni, O., Hadad, E., Kolodner, E.K., Moatti, Y., Lorenz, D.H.: Guaranteeing high availability goals for virtual machine placement. In: ICDCS, pp. 700–709 (2011)
17. Purshouse, R.C., Fleming, P.J.: On the evolutionary optimization of many conflicting objectives. In: TEVC, pp. 770–784 (2007)
18. Schroeder, B., Gibson, G.A.: A large-scale study of failures in high-performance computing systems. In: TDSC, pp. 337–351 (2010)
19. Google/roadef/euro challenge 2012, http://challenge.roadef.org/2012/en/
20. Voorsluys, W., Broberg, J., Venugopal, S., Buyya, R.: Cost of virtual machine live migration in clouds: A performance evaluation. In: Jaatun, M.G., Zhao, G., Rong, C. (eds.) Cloud Computing. LNCS, vol. 5931, pp. 254–265. Springer, Heidelberg (2009)
21. Filani, D., He, J., Gao, S., Rajappa, M., Kumar, A., Shah, P., Nagappan, R.: Comparing vm-placement algorithms for on-demand clouds. In: Dynamic Data Center Power Management: Trends, Issues, and Solutions (2008)
22. Datacentre energy efficiency, http://re.jrc.ec.europa.eu/energyefficiency/html/standby_initiative.htm
23. Xu, J., Fortes, J.A.: Multi-objective virtual machine placement in virtualized data center environments. In: GreenCom, pp. 179–188 (2010)
24. Lien, C.-H., Bai, Y.-W., Lin, M.-B.: Estimation by software for the power consumption of streaming-media servers. In: TIM, pp. 1859–1870 (2007)
25. Gandibleux, X., Martin, B., Perederieieva, O., Rosembly, S.: Sur la résolution approchée en trois étapes du sac-à-dos bi-objectif unidimensionnel en variables binaires. In: ROADEF, pp. 2–4 (2011)
26. Falkenauer, E.: Genetic algorithms and grouping problems (1998)
27. Zitzler, E., Laumanns, M., Thiele, L., Fonseca, C.M., da Fonseca, V.G.: Why quality assessment of multiobjective optimizers is difficult. In: GECCO, pp. 666–673 (2002)
28. Zitzler, E., Thiele, L., Laumanns, M., Fonseca, C.M., Da Fonseca, V.G.: Performance assessment of multiobjective optimizers: An analysis and review. In: TEVC, pp. 117–132 (2003)
29. Fleischer, M.: The measure of pareto optima applications to multi-objective metaheuristics. In: Fonseca, C.M., Fleming, P.J., Zitzler, E., Deb, K., Thiele, L. (eds.) EMO 2003. LNCS, vol. 2632, pp. 519–533. Springer, Heidelberg (2003)
30. Panigrahy, R., Talwar, K., Uyeda, L., Wieder, U.: Heuristics for vector bin packing. Research. Microsoft. Com (2011)

Hybrids of Integer Programming and ACO for Resource Constrained Job Scheduling

Dhananjay Thiruvady, Gaurav Singh, and Andreas T. Ernst

CSIRO, Australia

Abstract. A recent line of research considers hybrids of Lagrangian relaxation and Ant Colony Optimisation (ACO). Studies have shown that for hard constrained optimisation problems Lagrangian relaxation can effectively guide ACO to provide good feasible solutions. We consider applying these ideas to create a matheuristic combining ACO with decomposition approaches from mathematical programming for a resource constrained job scheduling problem. We are given a number of jobs which have to be executed on a number of machines satisfying several constraints. These include precedences and release times within machines and the machines are linked via a central resource constraint. By removing the linking constraint, the each machine's scheduling problem can be solved independently as a relatively simple subproblem. Both Danzig-Wolfe decomposition with column generation and Lagrangian relaxation are tried to carry out this decomposition. The relaxed solutions can provide useful guidance to determine solutions either via problem specific heuristics and ACO. Empirical results show that the Lagrangian relaxation matheuristic performs well in limited time-frames whereas the column generation based heuristic provides improved lower and upper bounds when run to convergence.

1 Introduction

We consider a problem motivated by an application in mining supply chains. Minerals from various mining sites have to be transferred by rail or road to ports and the resources available on these modes of transport such rail wagons and trucks are limited and have to be shared. This is due to the remote locations of the mines where limited infrastructure is available.

This problem can be formulated as resource constrained job scheduling (RCJS) problem. The transport of minerals can be viewed as jobs which must be executed on machines or mines. Each job requires some amount of the available resource and the cumulative resource requirements of jobs executing at the same time must not exceed the available resource. In some cases, certain materials may have to arrive at the ports before others which amount to enforcing precedences between jobs. Furthermore, certain materials have to arrive at the ports by specific times imposing due times on the jobs. If these due times are not satisfied, ships at the ports waiting for the materials may have to pay significant demurrage costs. Thus, the objective of this problem is to minimise the total

M.J. Blesa, C. Blum, and S. Voß (Eds.): HM 2014, LNCS 8457, pp. 130–144, 2014.

weighted tardiness. This problem has been previously considered in [17], where a Lagrangian relaxation heuristic is developed for this problem. A variant of the problem with hard deadlines is considered in [22], where it is shown that a hybrid constraint programming and ant colony optimisation approach is effective for this problem.

A similar class of problems with shared resources is project scheduling[1], with an extensive literature (eg [1,8,9,16]). Branch & bound and heuristic/local search approaches are investigated in [8]. Demeulemeester & Herroelen (2002) [9] also consider a variety of project scheduling problems and discuss exact approaches and meta-heuristics such as simulated annealing and genetic algorithms. Neumann *et al.* (2003) [16] consider a variant with time windows and also investigate various exact and heuristic approaches. Ballestin & Trautmann (2008) [1] investigate a project scheduling problem similar to the RCJS where the objective is to minimise the deviation from the completion times across all tasks. They show that a population-based iterated local search is effective on this problem. Regarding the objective, some studies have also considered total weighted tardiness [18,19]. These two studies consider similar problems but assume that the data is decentralised. They use an agent-based approach to determine good schedules in the presence of minimal information sharing.

Two popular methods within the mixed integer programming community are lagrangian relaxation (LR) [24,12] and column generation (CG) [3,2,24]. In this paper we are particularly interested in Danzig-Wolfe decomposition based column generation approaches. Here a restricted master problem is used to determine dual prices which then feed into the costs for one or more subproblems. Hence LR and CG can be viewed as closely related: underlying both of these methods is the idea of relaxing a number of complicating constraints, determining appropriate penalties for violating these constraints and building feasible solutions by using the relaxed solution as guidance. the main distinction is that in LR there is no master-problem, instead sub-gradient optimisation [5] is used to determine appropriate lagrangian or dual values. For the RCJS problem discussed above, LR is already proven [17] though not in combination with ACO. To the authors' knowledge CG has not previously been applied to this problem.

Meta-heuristics provide an alternative to deal with scheduling and combinatorial optimisation problems [6]. Within this class of algorithms, ACO has been demonstrated to work well on a range of problems [10,11]. It is based on the foraging behaviour of real ants when they go in search of food. On the paths they use to the food sources, the ants deposit a chemical (pheromone) which provide guidance on how good the food sources were. In the future, when other ants go out looking for food, they make use of the amount of pheromone to determine which paths to take. The paths with more pheromones receive more ants, who will in turn, deposit more pheromones. This feedback loop results in favouring better paths and eventually the colony converges to one of the best paths.

In the context of the RCJS problem, we consider hybrids of LR and ACO and CG and ACO. Meta-heuristics like ACO, often struggle when dealing with hard

---

[1] See [8] for an overview of the variants of project scheduling problems.

constraints but are very good at exploring search spaces whereas LR and CG provide a way of intelligently dealing with constraints or penalising violation. Furthermore the methods are complementary in that ACO only provides heuristic solutions (upper bounds) while LR & CG primarily provide lower bounds and often don't find good upper bounds without the use of additional techniques (such as repair heuristics or branch and bound). Thus, by combining these approaches, we aim to determine good feasible solutions to the RCJS problem in a reasonable time-frame while also getting some indication of solution quality via lower bounds. Hybrids of LR and metaheuristics have already been explored [7]. Furthermore, LR and ACO hybrids have been applied to project scheduling [23] and car sequencing [21]. CG and ACO combinations have also been explored with vehicle routing [13,14].

The paper is organized as follows. The problem is formally described next in Section 2. The following sections, Section 3, 5, 4 and 6, discuss the details of all the methods implemented. This is followed by the experiments and results in Section 7. The paper concludes with Section 8.

2   Problem Specification

Formally, the RCJS problem can be defined as follows. We are given a number of jobs $\mathcal{J} = \{1, \ldots, n\}$ to be processed on machines $\mathcal{M} = \{1, \ldots, l\}$. Each job i has a release time r_i, processing time p_i, due time d_i, weight w_i, resource requirement g_i and a machine m_i. A job must execute on the machine it is assigned to and a machine may only execute one job at a time. Jobs assigned to the same machine may have precedences between them: for two jobs $i, j \in \mathcal{J}$, if i precedes j $(i \rightarrow j)$, then j may start only after i completes. The total amount of resource consumed by all jobs executing concurrently on the set of machines is limited to $\mathcal{G}$. While we could treat time as being continuous and with an infinte horizon, it is convenient for the presentation here to consider discrete time with a finite horizon. In particular if all of the processing times and relase times are integer then only integer times need to be considered and we can easily compute a heuristic bound on the maximum completion time of any optimal schedule (eg $D = \max_{j \in \mathcal{J}} r_j + \sum_{j \in \mathcal{J}} p_j$). Hence we define the set of time intervals $\mathcal{T} = \{0, \ldots, D\}$.

The ACO component makes use of a sequence or permutation π of jobs. From π a feasible schedule $\mathcal{S}(\pi)$ may be obtained by assigning each job i in the order of the permutation the first start time s_i such that all of the constraints are satisfied. To be feasible we require that: $s_i \geq r_i$, $s_j \geq s_i + p_i$ if $i \rightarrow j$. Further, let P_t be the set of jobs either starting at time t or being processed at time t:

$$P_t = \{j | s_j \leq t < s_j + p_j, j \in \mathcal{J}\}. \tag{1}$$

Now $\mathcal{S}(\pi)$ is considered *resource feasible* if at any time t the amount of resource consumed by jobs across all machines does not exceed the capacity $\mathcal{G}$:

$$\sum_{j \in P_t} g_j \leq \mathcal{G} \qquad \forall t. \tag{2}$$

The objective is to minimize the total weighted-tardiness of a resource feasible schedule $\mathcal{S}(\pi)$

$$f(\mathcal{S}(\pi)) = \sum_{i=1}^{n} w_i \times T_i(s_i) \qquad (3)$$

where $T_i(s_i)$ is the tardiness of the job i and can be defined as $\max(s_i + p_i - d_i, 0)$.

In many of the optimisation methods presented in this paper we make use of Algorithm 1 to create a resource feasible schedule. In this algorithm jobs may not be scheduled in the order in which they appear in the sequence (ie it is possible that $s_{\pi_{i+1}} < s_{\pi_i}$). This is because a subsequent job on a different machine may be able to run earlier if it requires less resource. Also for any problem an optimal solution can be constructed using Algorithm 1 using some sequence π^*. In fact if we knew an optimal schedule s^* we could construct such a sequence π^* by sorting jobs in order of their start times s_j^* and Algorithm 1 would return a set of start times with $s_j \leq s_j^*$.

Algorithm 1. Schedule a Sequence of Jobs

Require: A RCJS instance and a permutation of the jobs π
Ensure: A resource feasible schedule $\mathcal{S}(\pi)$ defined by job start times $s_j \; \forall \, j \in \mathcal{J}$
 1: **for all** $i < j$ **do**
 2:     **if** $\pi_j \to \pi_i$ **then** modify π by inserting π_i in position $j + 1$
 3: **end for**
 4: $R_t := \mathcal{G} \; \forall \, t \in \mathcal{T}$
 5: **for** $i = 1, \ldots, n$ **do**
 6:     $t^{\min} := \max \left\{ r_{\pi_i}, \max \left\{ s_{\pi_j} + p_{\pi_j} \mid j < i \wedge m_{\pi_i} = m_{\pi_j} \right\} \right\}$
 7:     $s_{\pi_i} := \min \left\{ t \geq t^{\min} \mid R_{t+k} \geq g_{\pi_i} \; \forall \, 0 \leq k < p_{\pi_i} \right\}$
 8:     $R_t := R_t - g_{\pi_i} \; \forall \, t = s_{\pi_i}, \ldots, s_{\pi_i} + p_{\pi_i} - 1$
 9: **end for**
10: **return** $\mathcal{S}(\pi) = \{s_1, \ldots, s_n\}$

---

3   An Integer Programming Formulation

The problem can be formulated as an integer linear program (ILP) as follows. Let z_{jt} be a binary variable which is 1 iff job $j \in J$ is completed by time t or earlier.

$$\min \quad \sum_{j \in \mathcal{J}} \sum_{t \in \mathcal{T}} c_{jt} \left(z_{jt} - z_{j,t-1} \right) \qquad (4)$$

$$\text{s.t.} \quad \sum_{j \in \mathcal{J}^m} z_{j,t+p_j} - z_{jt} \leq 1 \qquad \forall \, t \in \mathcal{T}, \; \forall \, m \in M \qquad (5)$$

$$z_{jt} \geq z_{j,t-1} \qquad \forall \, j \in J, \; \forall \, t > 0 \qquad (6)$$

$$z_{jT} = 1 \qquad \forall \, j \in J \qquad (7)$$

$$z_{j,t-p_k} \geq z_{kt} \qquad \forall \, j \to k, \; \forall \, t \in \mathcal{T} \qquad (8)$$

$$z_{j,t} = 0 \qquad \forall\, t \in 0, \ldots, r_j + p_j - 1,\ \forall\, j \in J \qquad (9)$$

$$\sum_{j \in J} g_j \left(z_{j,t+p_j} - z_{jt} \right) \leq \mathcal{G} \qquad \forall\, t \in \mathcal{T} \qquad (10)$$

$$z_{jt} \in \{0, 1\} \forall\, t \in \mathcal{T}$$

where $c_{jt} = w_j \max\{t - d_j, 0\}$, $\forall j \in J, \forall\, t \in \mathcal{T}$. Equation (4) is the objective. Equation (5) requires that only one job on a machine is processed at one time. Equation (6) requires that a job stays completed once it has been finished. Equation (7) requires that all jobs are completed and Equation (8) specifies the constraints for the precedences. Equations (9) enforces the release times and finally, Equation (10) is the resource constraint across all machines.

4   Column Generation

Danzig-Wolfe decomposition can be used to reformulate the above problem in a way that separates the scheduling subproblem for each machine. Instead of the original z variables we define x_{mc} if machine $m \in \mathcal{M}$ uses schedule c. In principle we would need to enumerate all of the possible schedules c (or at least all extreme points of the polyhedron associated with the scheduling problem for machine m). However in practice it is sufficient to simply use a pool of such schedules or *columns* which is iteratively updated. This gives rise to a Restricted Master Problem (RMP) and a set of column generation or pricing subproblems for each machine (see [24]). Let $\mathcal{C}_m$ be the set of schedules for machine m, $\mathcal{C} = \uplus_m \mathcal{C}_m$ then:

Problem RMP(C)

$$\min \sum_{m \in \mathcal{M}} \sum_{c \in \mathcal{C}_m} f_{mc}\, x_{mc} \qquad (11)$$

$$\text{s.t.} \qquad \sum_{c \in \mathcal{C}_m} x_{ic} = 1 \qquad \forall\, m \in \mathcal{M} \qquad (12)$$

$$\sum_{m \in \mathcal{M}} \sum_{c \in \mathcal{C}_m} s_{mct} \cdot x_{mc} \leq \mathcal{G} \qquad \forall t \in \mathcal{T} \qquad (13)$$

$$x_{mc} \in \{0, 1\} \qquad \forall\, m \in \mathcal{M},\ c \in \mathcal{C}_m$$

where f_{mc} is the weighted tardiness cost of schedule c on machine m. The first constraint ensures only one schedule per machine is selected. Constraint (13) is the maximum resource constraint and s_{mct} is the amount of resource used by the cth solution in the mth machine at time point t. In order to solve this problem exactly we would need to employ branch-and-price [2]. However here we are only interested in getting heuristic solutions so we solve the continuous relaxation of RMP (ie $0 \leq x_{mc} \leq 1$) and then use the dual prices λ_t of constraints (13) to generate new columns. The pricing problem for generating new schedules for each machine m is:

Algorithm 2. CG for RCJS

Require: An RCJS instance
1: $\pi^{bs} :=$ NULL (best solution)
2: initialize $\lambda_t^0 = 0, \forall t \in \mathcal{T}$
3: $\gamma := 2.0,\ k := 0,\ gap := \infty,\ UB^* := \infty,\ LB^* := 0$
4: $\mathcal{C} =$ generateCols()
5: **while** $\gamma > 0.01$ & $\frac{UB^* - LB^*}{UB^*} > 0.01$ & $sf > 0.01$ **do**
6:     **for all** $m \in M$ **do**
7:         Solve $SP_m(\lambda)$ to obtain $L_m(\lambda)$ and a new column c_m
8:         $\mathcal{C}_m := \mathcal{C} \cup \{c_m\}$
9:     **end for**
10:    $LB^* := \max\{LB^*, L(\lambda)\}$
11:    $\pi :=$ GenerateSequence$(c_1, \ldots, c_{|\mathcal{M}|})$
12:    ImproveUB(π)
13:    UpdateBest(π^{bs}, π, γ)
14:    $UB* := f(\pi^{bs})$
15:    Solve RMP$(\mathcal{C})$ to obtain objective MLB
16:    $sf := \frac{MLB - LB^*}{MLB}$
17:    $\lambda :=$ UpdateDual(λ)
18: **end while**
19: **return** π^{bs}

Problem SP$_m(\lambda)$

$$L_m(\lambda) = \min_z \sum_{j \in \mathcal{J}^m} \sum_{t>0} \left(c_{jt} + \sum_{i=1}^{\min\{p_j, t\}} \lambda_{t-i}\, g_j \right) (z_{jt} - z_{j,t-1}) \tag{14}$$

$$\text{subject to (5)-(8).}$$

This problem is still not trivial to solve but due to its smaller size and simpler structure compared to the original problem, it can usually be optimised using a commercial MILP solver in a relatively short amount of time. Note that for any $\lambda \geq 0$ a lower bound for problem may be calculated using the lagrangian function:

$$L(\lambda) = \sum_{m \in \mathcal{M}} L_m(\lambda) - \mathcal{G} \sum_{t \in \mathcal{T}} \lambda_t \tag{15}$$

The Column Generation (CG) algorithm is presented in Algorithm 2. In line 4 a number of feasible columns are generated. This essentially involves determining the start times for tasks on each machine such that a resource feasible solution may be found. This will ensure the RMP may at least generate one solution to begin with. Within the main loop (line 5) the column generation procedure is executed. First the sub-problems for each machine are solved resulting in the lower bound $L(\lambda)$. If new columns are found for any machine they are added to the pool of solutions (line 8). We can also generate a permutation of the jobs π by sorting them in the order of the start times produced by the subproblems (line 11).

The sequence π can be used to generate a feasible solution (see Algorithm 1) and this sequence is improved using an ACO heuristic as described in Section 6.

At Line 15 the LP relaxation of the master problem RMP is solved. We could now update the dual values simply by using the optimal dual values λ_t^* of LP solution. However this tends to be fairly unstable. Here we update λ using

$$\lambda_t := 0.5\,\lambda_t^* + 0.5\,\hat{\lambda}_t \qquad \forall\, t \in \mathcal{T} \tag{16}$$

where $\hat{\lambda}_t$ are the dual prices that resulted in the highest lower bound at Line 10. This linear combination allows for stabilisation[2] of the dual prices and serves a similar purpose as bundle methods [4].

The objective value of $RMP(\mathcal{C})$ is an upper bound on the original lower bound. This is used to compute the stability factor sf in Line 16. A low stability factor implies that the lower bound has reached its limit and thus the algorithm can terminate. Alternatively we could continue until no new columns with negative reduced cost are generated in the pricing subproblems

5   Lagrangian Relaxation

A related decomposition method to Danzig-Wolfe decomposition is Lagrangian relaxation. By relaxing the resource constraints (9) using Lagrangian multipliers $\lambda_t \geq 0$, $\forall t \in \mathcal{T}$, we obtain the lagrangian function $L(\lambda)$ provided in (15). This again results in the same subproblems as we have seen in the column generation approach. However instead of solving a restricted master problem we attempt to solve the Lagrangian optimisation problem

$$\max_{\lambda \geq 0} L(\lambda) \tag{17}$$

As this is a non-smooth optimisation problem, subgradient optimisation is commonly used to search for a good value of λ.

5.1   The Lagrangian Heuristic

A Lagrangian heuristic can be defined as shown in Algorithm 3. This algorithm is based loosely on previous work on Lagrangian relaxation for the RCJS in [17] but differs in the use of ACO to obtain better solutions as well as some of the details of the implementation. After initialising the multipliers and various parameters, the main loop starts at line 4 and executes for 1000 iterations or while the gap and γ are above a threshold.

The relaxed problem is solved in line 5 and we extract both the objective function $L_m(\lambda)$ and schedule $z(\lambda)$. After all machine subproblems are solved, a complete set of start times are available, but possibly violating the resource constraint. In order to determine a feasible solution from these start times a

---

[2] We found that this simple linear combination is effective. However, more complex stabilisation schemes may be attempted [15], also including bundle methods [4].

Algorithm 3. LR for RCJS

Require: A RCJS instance
1: $\pi^{bs} := $ NULL (best solution)
2: initialize $\lambda_t^0 = 0, \forall\, t \in \{1, \ldots, D\}$
3: $\gamma := 2.0$, $k := 0$, $UB^* := \infty$, $LB^* := 0$, $i := 0$
4: **while** $\gamma > 0.01$ & $\frac{UB^* - LB^*}{UB^*} > 0.01$ & $i < 1000$ **do**
5:   **for all** $m \in \mathcal{M}$ **do** Solve $SP_m(\lambda)$ to obtain $L_m(\lambda^i)$ and $z(\lambda^i)$
6:     $LB^* := \min \left\{ LB^*, \sum_m LRR(\lambda^i)^m - \mathcal{G}\sum_t \lambda_t^i \right\}$
7:     $\pi := $ GenerateSequence($z(\lambda^i)$)
8:     ImproveUB(π)
9:     UpdateBest(π^{bs}, π, γ)
10:    $UB^* := f(\pi^{bs})$
11:    $\lambda^{i+1} := $ UpdateMult(λ^i, LB^*, UB^*, $z(\lambda^i)$, γ)
12:    $i := i + 1$
13: **end while**
14: return π^{bs}

repair heuristic is used as before. That is we apply first Algorithm 1 to the sequence of jobs sorted by start times in the subproblems. Then ImproveUB(π) uses ACO (see Section 6) to improve this sequence.

UpdateBest(π^{bs}, π, γ) does two things: $\pi^{bs} := \pi$ and $\gamma = \min\{1.3\gamma, 2.0\}$ if $f(\pi) < f(\pi^{bs})$, and $\gamma := 0.95\gamma$ otherwise. The next procedure is UpdateMult(λ^i, LB^*, UB^*, x, γ) which makes use of subgradient optimisation [5] to update the Lagrangian multipliers $t \in \mathcal{T}$:

$$\lambda_t^{i+1} = \max\left\{ 0, \lambda_t^i + \frac{\gamma(UB^* - LB^*)}{\sum_{\hat{t}\in\mathcal{T}} \Delta_{\hat{t}}^2} \Delta_t \right\} \tag{18}$$

where $\Delta_t = (1.0 - \phi)\hat{\Delta}_t + \phi(\mathcal{G} - \sum_{j\in J} g_j(z_{t,t+p_j} - z_{tj}))$. $\hat{\Delta}$ are the best set of multipliers known, for further details we refer the reader to [17].

6 Ant Colony Optimisation

The ACO algorithm used in the hybrid matheuristic is based on the implementation of ACO for the RCJS discussed in [20]. For the sake of completeness we provide the details here. The variant of ACO used here is sometimes referred to as ant colony system (ACS) [11] and is presented in Algorithm 4. The pheromones τ_{ij} represent the desirability of picking job j in position i, i.e., $\pi_i = j$.

A complete sequence of jobs is obtained by incrementally adding jobs to π (ConstructSequence()). The selection of a job is done in one of two ways. A random number $q \in (0, 1]$ is generated and compared to a pre-defined parameter q_0. If $q < q_0$, a deterministic selection is used to pick job k for variable i according to

$$k = \operatorname*{argmax}_{j \in \mathcal{J}} \tau_{ij} \tag{19}$$

Algorithm 4. ACO for RCJS

Require: A RCJS instance, $\mathcal{T}$, initial solution π^{bs}
Ensure: Updated solution π^{bs} that is at least as good as the input solution.
1: **while** termination conditions not satisfied **do**
2:    $S := \emptyset$
3:    **for** $j = 1$ to n_{ants} **do** $\pi^j :=$ ConstructSequence()
4:    $\pi^{ib} := \arg\min_{j=1,\ldots,n_{ants}} f(\pi^j)$
5:    $\pi^{ib} :=$ Improve(π^{ib})
6:    $\pi^{bs} :=$ Update(π^{ib})
7:    $\mathcal{T} :=$ PheromoneUpdate(π^{bs})
8: **end while**
9: **return** π^{bs}

Otherwise, a probabilistic selection is used

$$P(\pi_i = k) = \frac{\tau_{ik}}{\sum_{j \in \mathcal{J}} \tau_{ij}} \tag{20}$$

The solutions are built in Line 3 and then the iteration best (π^{ib}) is improved with local search ($\pi^{ib} :=$ Improve(π^{ib})).[3] The best solution found so far π^{bs} is updated in line 6 if π^{ib} is an improvement. This is followed by an update to the pheromone trails based on the solution components in π^{bs} ($\mathcal{T} :=$ PheromoneUpdate(π^{bs})) using the following formula:

$$\tau_{ij} = \tau_{ij} \cdot (1.0 - \rho) + \delta \tag{21}$$

where $\delta = Q/f(\pi^{bs})$, and the reward Q is selected such that $0.01 \leq \delta \leq 0.1$. The evaporation rate $\rho = 0.1$ was determined from [22].

6.1   Variants of CG/LR and ACO Hybrids

There are two different ways of integrating ACO with the CG or LR algorithms. As we have described so far, the ACO algorithm can perform a few iterations to try to improve the best upper bound in each main iteration of the CG or LR algorithm. That is ACO is run as part of ImproveUB(π) in Algorithm 2 or Algorithm 3. This has the advantage that if the ACO finds new solutions these can be used by the mathprogramming methods. For the CG method the ACO provides additional columns to be used in the RMP problem. Hence for CG the ACO provides a way of heuristically expanding and diversifying the set of columns. For LR the feedback is less immediate, though having better upper bounds can improve the step-size calculation in Equation (18). The downside of this approach is that the ACO can consume a non-trivial amount of computational time and spend much of this time looking at solutions far from the optimum. We refer to this integrated version as CG-ACO and LR-ACO.

---

[3] See [22] for beta-sampling which is used as the improvement method here.

An alternative is to run the ACO for a longer time but only after the main loop of the mathprogramming algorithm has finished. This second type of hybrid will be referred to as CG+ACO and LR+ACO respectively. In order to learn from the effort during the main mathprogramming loop and to warm-start the ACO, a partial converged pheromone matrix is maintained through the CG or LR run. The pheromones are intialised: $\tau_{tj} = 1.0 \ \forall \ t \in \mathcal{T}$ and $j \in \mathcal{J}$. After every LR or CG sub-problem solve, the repaired solution π produced by GenerateSequence is determined and corresponding to its solution components, the pheromones are updated $\forall i, j \in \pi$: $\tau_{ij} \leftarrow \tau_{ij} + 1.0$. Over a number of iterations, this scheme essentially favours the relaxed solutions seen so far. This scheme also requires relatively few ACO iterations since the pheromone matrix has converged. Similar ideas have been tried on other combinatorial optimisation problems in [23,21] though only in the context of Lagrangian relaxation

7 Experiments and Results

Problem instances from [17][4] were considered for the experiments, with the same subset used as in [20]. Additionally, a number of larger instances were generated with up to 200 jobs to determine how the algorithms scale with larger problem sizes. CG and LR were the two base algorithms considered here using either interleaved (CG-ACO, LR-ACO) or sequential (CG+ACO, LR+ACO) integration with ACO.

Thirty runs per instances were conducted on two Xeon X7350 machines at 2.93GHz with 65 GB of shared memory. The machines consist of 4 quad-core CPUs allowing 16 parallel threads. The sub-problems were solved with CPLEX 12.5. For the implementation reported here only multiple threads were only used within the CPLEX library to solve the $SP(\lambda)$ subproblems.

The parameters for the ACO component were set according to [20] and the LR and CG parameters were determined from [17] and through tuning by hand. n_a was set to 10 as the number of solutions to construct per iteration of ACO. The learning rate was selected to be relatively high, $\rho = 0.1$ and $q_0 = 0.9$ was chosen to favour high deterministic selection. For the hybrids, 500 ACO iterations were chosen for LR-ACO and CG-ACO as means to find improvements quickly and also since there will be multiple ACOs that are run over the course of the algorithm's execution. For LR+ACO and CG+ACO, a larger number of ACO iterations are conducted (5000) since ACO is run only once for these implementations.

Two sets of experiments were conducted. In the first experiment, each algorithm was given a time limit of 3 hours based on CPU cycles. In the second experiment, all the algorithms were run for 100 iterations to allow them to converge substantially.

---

[4] Data available from http://dx.doi.org/10.4225/08/506B728AB6308

7.1   Results

Table 1 shows results for all algorithms for 3 hours of CPU time. Overall, we see that ACO provides advantages for the small and medium size problems up to 12 machines. For these problems, we also see that CG-ACO is nearly always better than the other two CG-based algorithms. This shows that the additional columns provided by ACO to CG assist with convergence. Beyond 12 machines, the LR-based algorithms are more effective with all CG-based algorithms being outperformed. On close investigation we find that the number of iterations conducted by both algorithms is very different with the CG-based algorithms completing only about half the iterations compared to the LR-based algorithms for the large instances. For example, for the instance 20-2, CG does 15 iterations compared to LR which completes 31 iterations within the 3 hour CPU time limit. This is a large difference which accounts for the poor performance of the CG-based algorithms for the large instances.

Given that the sub-problems in LR or CG are similar, we would expect both implementations to carry out a similar number of iterations in the same time-frame. It is possible that there may be a difference of one or two iterations in favour of LR since CG solves the master problem. However, the differences seen are much greater and we determined two reasons for observing this effect. Firstly, the range on dual prices associated with CG are much larger than LR. This makes the CG subproblem slower to solve. Secondly, most of the Lagrangian multipliers, at least in the early iterations, are nearly all zero whereas the the dual prices are often positive-valued. This also leads to slower sub-problem solving time for CG.

Considering the upper bounds, we see that ACO clearly provides improvements throughout whether it is combined with LR or CG for up to 15 machines. This is not surprising since ACO is designed to improve the upper bounds. However, the CG and ACO hybrids are the most effective across all the instances. This implies that the dual prices provide more useful information than the Lagrangian multipliers when converting the relaxed solutions to their feasible counterparts. Beyond 12 machines, LR performs best which is attributable to the larger number of sub-problems solved by this method.

Considering the lower bounds, CG, CG-ACO and LR are all effective. In theory all of these methods should produce similar lower bounds with minor differences due to time spent on the ACO component for example. Hence, differences are purely due to the rate of convergence for the different approaches. Overall, CG-ACO is the most effective up to 12 machines, beyond which LR is more effective. Clearly, the additional columns provided by ACO assist CG to even generate improved lower bounds. Thus for CG, ACO proves advantages in two different ways. As mentioned earlier, there are very few iterations conducted with CG for instances with 15 machines and hence, LR is significantly better here. The LR and ACO hybrids are not as effective since the ACO component does not provide any feedback via the multipliers.

Table 1. Results of all algorithms on all instances for 3 hours of CPU time. The upper bounds are reported as the % to the best upper bound (UB*) found for each instance (listed in column 2). The lower bounds are reported as $100 - \%$ to the best upper bound (UB*) found for each instance. Standard deviations are provided within parentheses (·) and are appropriately scaled.

Instance	UB*	CG LB	CG UB	CG-ACO LB	CG-ACO UB	CG+ACO LB	CG+ACO UB	LR LB	LR UB	LR-ACO LB	LR-ACO UB	LR+ACO LB	LR+ACO UB
3-5	505	3.8(0.1)	15.0(3.6)	**3.8(0.1)**	5.7(2.3)	3.8(0.1)	5.9(4.1)	3.9(0.0)	17.7(0.2)	4.0(0.0)	*1.7(1.4)*	3.9(0.0)	6.2(2.3)
3-23	149.07	1.1(0.1)	4.1(1.5)	1.2(0.0)	1.1(0.8)	1.2(0.1)	1.2(0.7)	**1.0(0.0)**	3.8(0.0)	1.2(0.0)	*0.0(0.0)*	**1.0(0.0)**	1.0(0.8)
3-53	69.36	6.6(0.1)	7.4(3.5)	**6.6(0.1)**	0.7(0.5)	6.6(0.1)	1.0(0.4)	**7.8(0.2)**	9.4(3.2)	7.8(0.2)	*0.0(0.0)*	7.9(0.1)	1.2(1.2)
4-28	23.81	7.8(0.0)	28.2(12.4)	7.8(0.0)	1.5(1.5)	7.8(0.0)	3.4(4.1)	7.8(0.0)	21.8(5.8)	7.8(0.0)	*0.2(0.2)*	7.8(0.0)	2.1(2.5)
4-42	66.73	**33.3(0.0)**	59.3(11.4)	**33.3(0.0)**	4.4(3.1)	**33.3(0.0)**	7.5(4.6)	35.1(0.4)	105.1(23.5)	35.0(0.4)	*2.0(0.7)*	35.2(0.5)	14.4(7.1)
4-61	45.96	10.9(0.1)	25.5(9.7)	10.9(0.1)	0.4(1.0)	10.8(0.0)	2.7(3.9)	11.1(0.1)	9.5(4.4)	13.5(0.6)	*0.0(0.0)*	**10.7(0.1)**	1.2(0.9)
5-7	252.9	21.6(0.1)	19.9(3.9)	21.6(0.0)	2.8(1.4)	21.6(0.0)	4.5(3.0)	23.5(0.3)	23.6(4.3)	23.6(4.3)	*2.4(1.3)*	23.4(0.2)	6.4(4.2)
5-21	168.63	**5.9(0.0)**	37.7(8.3)	5.9(0.0)	3.7(1.7)	5.9(0.1)	4.9(3.1)	6.1(0.0)	46.3(3.9)	6.1(0.0)	*0.4(0.5)*	6.1(0.1)	5.2(4.2)
5-62	250.67	19.4(0.1)	22.2(3.6)	**19.3(0.0)**	9.6(2.5)	19.4(0.0)	9.7(4.2)	21.1(1.3)	28.3(11.7)	21.5(1.3)	10.0(3.5)	21.0(1.0)	**7.6(4.5)**
6-10	861.35	**18.0(0.1)**	11.4(2.2)	**18.0(0.1)**	10.6(1.3)	18.1(0.0)	**3.7(2.0)**	18.4(0.1)	20.4(3.8)	18.5(0.1)	13.0(1.3)	18.4(0.1)	7.3(2.8)
6-28	228.46	25.8(0.1)	18.3(3.2)	**25.8(0.1)**	5.9(2.1)	25.9(0.0)	**4.9(2.8)**	29.2(0.6)	24.4(4.5)	29.3(0.4)	5.5(2.1)	29.0(0.6)	7.2(3.8)
6-58	243.2	**19.0(0.1)**	18.8(3.8)	19.0(0.1)	6.6(1.6)	19.1(0.0)	**5.7(3.4)**	20.8(1.1)	33.3(13.2)	21.4(1.0)	9.6(2.4)	21.2(0.9)	12.8(6.2)
7-5	438.71	**18.2(0.1)**	15.2(3.5)	18.2(0.1)	7.7(2.5)	18.4(0.0)	**4.4(3.0)**	19.1(0.1)	26.2(6.6)	19.5(0.5)	10.9(3.6)	19.2(0.4)	6.3(3.6)
7-23	562.82	21.1(0.1)	14.5(2.3)	**21.1(0.1)**	9.8(1.5)	21.2(0.0)	**5.4(2.1)**	22.4(0.6)	24.8(4.8)	22.6(0.6)	12.9(3.1)	22.5(0.5)	8.9(3.8)
7-47	439.41	25.9(0.1)	19.0(3.5)	**25.9(0.1)**	14.4(3.6)	26.1(0.0)	**7.5(4.2)**	27.2(0.6)	27.0(4.5)	27.4(0.6)	16.3(2.6)	27.0(0.3)	10.5(3.1)
8-3	631.81	**21.8(0.1)**	19.9(2.4)	21.8(0.0)	19.6(2.7)	21.9(0.0)	**9.5(3.0)**	22.0(0.1)	24.2(2.5)	22.0(0.0)	26.4(2.9)	22.0(0.1)	12.4(4.7)
8-53	449.22	16.7(0.1)	19.2(2.5)	16.7(0.0)	10.7(2.2)	16.8(0.0)	**6.2(2.8)**	18.2(0.5)	31.7(3.9)	18.4(0.3)	13.4(2.6)	18.2(0.5)	9.8(3.2)
8-77	1237.21	19.2(0.5)	11.6(1.8)	**17.6(0.1)**	6.2(1.1)	17.9(0.3)	**3.6(2.2)**	18.4(0.1)	16.3(1.5)	19.2(0.1)	10.4(1.5)	18.4(0.1)	6.1(2.4)
9-20	930.18	**17.4(0.1)**	8.9(2.1)	17.4(0.1)	6.5(0.7)	17.5(0.0)	3.7(1.7)	17.6(0.2)	9.5(2.6)	17.7(0.3)	6.5(1.1)	17.6(0.2)	*3.5(1.7)*
9-47	1233.13	19.2(0.3)	7.6(1.2)	**18.6(0.0)**	5.5(1.0)	19.1(1.5)	*3.0(1.6)*	19.4(0.2)	11.3(1.8)	19.9(0.1)	10.4(1.1)	19.3(0.1)	5.9(2.0)
9-62	1460.72	16.7(0.2)	6.9(1.1)	16.6(0.1)	4.2(0.9)	16.7(0.0)	3.1(1.5)	**16.5(0.0)**	6.5(1.0)	16.6(0.0)	4.3(0.8)	**16.5(0.0)**	*2.4(1.3)*
10-7	2538.17	19.9(1.7)	7.8(1.9)	16.9(0.1)	4.9(1.6)	17.7(1.4)	*2.9(1.1)*	**16.9(0.1)**	7.2(1.3)	17.0(0.0)	6.4(1.2)	**16.9(0.1)**	3.8(1.6)
10-13	2191.75	18.8(2.6)	7.6(1.8)	16.2(0.1)	4.6(0.8)	16.4(0.6)	*3.0(1.4)*	16.2(0.0)	8.8(1.1)	16.4(0.1)	7.6(0.6)	**16.1(0.0)**	4.1(1.7)
10-31	614.57	20.9(0.1)	8.3(1.9)	**20.8(0.1)**	5.8(1.3)	20.9(0.0)	3.4(2.2)	21.1(0.1)	13.0(1.9)	21.1(0.1)	8.2(1.8)	21.1(0.1)	5.2(2.4)
11-21	1017.13	17.0(0.1)	7.0(1.3)	**17.0(0.1)**	3.9(1.1)	17.1(0.0)	2.4(1.5)	17.2(0.1)	9.0(1.9)	17.2(0.1)	6.2(1.0)	17.1(0.1)	3.2(1.4)
11-56	1790.09	18.3(0.7)	4.8(1.1)	**16.4(0.1)**	3.3(0.9)	16.8(0.3)	*1.9(1.1)*	16.8(0.0)	7.3(1.9)	17.3(0.1)	9.1(1.5)	16.8(0.1)	4.1(1.0)
11-63	2021.89	15.3(0.1)	4.3(1.0)	15.3(0.1)	3.1(0.8)	15.3(0.0)	*1.8(1.0)*	15.2(0.0)	5.3(1.0)	15.3(0.0)	4.9(1.0)	**15.2(0.1)**	2.7(1.4)
12-14	1766.43	18.2(0.4)	4.2(1.1)	**17.3(0.0)**	2.9(1.0)	17.5(0.3)	*2.1(0.8)*	17.9(0.0)	7.8(1.1)	18.1(0.0)	7.4(1.1)	17.8(0.1)	5.5(1.4)
12-36	2968.87	38.2(8.0)	12.6(3.3)	18.2(0.9)	*1.7(0.8)*	61.0(24.4)	5.7(2.8)	17.0(0.1)	6.5(0.9)	17.4(0.1)	7.1(1.0)	**16.9(0.1)**	4.2(1.6)
12-80	2457.55	69.8(6.6)	11.2(3.3)	24.9(1.1)	*2.3(1.3)*	47.5(19.4)	5.0(3.3)	20.6(0.0)	8.3(1.2)	21.3(0.1)	9.4(0.9)	**20.4(0.1)**	5.1(1.9)
15-2	3927.99	91.4(1.3)	25.2(3.3)	57.1(7.2)	9.4(2.5)	89.2(3.4)	12.3(3.2)	17.9(0.1)	3.4(0.8)	18.4(0.1)	3.8(0.9)	**17.6(0.2)**	*2.4(1.0)*
15-3	4327.26	90.9(0.0)	22.4(3.6)	25.7(1.1)	*2.5(1.3)*	85.4(5.6)	12.5(4.0)	17.5(0.1)	4.3(1.1)	17.8(0.1)	5.3(1.0)	**17.3(0.2)**	4.3(1.2)
15-5	3490.2	84.2(10.7)	13.5(3.1)	**16.3(0.2)**	*1.4(0.7)*	27.8(18.5)	3.2(2.7)	16.5(0.1)	4.7(0.8)	16.8(0.1)	5.1(0.9)	16.4(0.1)	3.8(0.9)
20-2	8344.87	90.5(3.8)	18.0(2.3)	**16.4(0.1)**	7.0(1.3)	88.6(4.8)	13.3(2.0)	**16.4(0.1)**	*2.3(0.5)*	16.9(0.1)	2.8(0.6)	16.7(0.9)	2.3(0.9)
20-5	14533.3	94.7(0.0)	48.7(7.4)	94.7(0.0)	29.6(3.99)	94.7(0.0)	28.3(4.4)	25.1(0.7)	1.4(0.5)	25.1(0.3)	1.9(0.7)	**24.7(1.8)**	*1.0(0.7)*
20-6	7438.49	95.9(0.5)	21.6(3.2)	53.5(5.8)	7.2(1.8)	95.7(1.3)	15.8(2.2)	**15.6(0.1)**	*1.6(0.8)*	18.1(0.6)	5.3(1.0)	15.6(0.1)	1.9(0.5)

Comparison by Iterations. The results above show that given a limited amount of CPU time, one of the LR and ACO hybrids is the best option. This is especially true for the larger instances. However, we often see that the CG-based algorithms do not converge in the given time-frame. Thus, we compare the algorithms on how they perform over 100 iterations and we only consider large instances ($\geq$ 10 machines). Note, the CG-based algorithms take significantly longer to complete the 100 iterations for these instances.

Table 2 compares the lower and upper bounds for 100 iterations and a subset of the instances. We first focus on the upper bounds. Apart from the first two instances with 10 machines, the CG-based algorithms provide the best upper bounds. This result shows that when CG is given sufficient time it can assist in providing better guidance with and without ACO compared to LR. Examining ACO's influence, we see that using it nearly always provide advantages to both LR and CG.

Table 2 also shows a comparison of the lower bounds for 100 iterations. Again, apart from a small number of instances (10-13,10-31,11-56 and 11-63), CG is always superior. The effect of ACO here is not as significant, especially with LR, compared to the upper bounds' result. This is not surprising since ACO is designed to improve the upper bounds and does not have a significant influence on the LR algorithms. However, with CG we see that although the lower bounds are always similar, CG-ACO does provide improvements on occasion (e.g. 12-36, 20-6). This can be attributed to the additional columns that ACO provide to CG to help it converge more quickly.

Table 2. Results of CG, CG-ACO, LR and LR-ACO on large instances for 100 iterations. The results are the % to the best upper bound (UB*) found in these 4 runs.

		CG		CG-ACO		LR		LR-ACO	
Instance	UB*	LB	UB	LB	UB	LB	UB	LB	UB
10-7	2655.33	0.794	0.016	0.794	0.002	0.793	0.041	**0.794** *0.000*	
10-13	2301.41	0.798	0.010	0.798	0.002	**0.798**	0.039	0.797 *0.000*	
10-31	652.79	0.744	0.000	**0.744** *0.000*		0.742	0.067	0.740	0.049
11-21	1066.3	0.791	0.031	**0.791** *0.000*		0.788	0.054	0.784	0.047
11-56	1839.05	0.812	0.007	0.812 *0.000*		**0.812**	0.056	0.811	0.021
11-63	2091.68	0.819	0.012	0.819 *0.000*		**0.820**	0.021	0.818	0.022
12-14	1838.08	**0.793**	0.001	0.793 *0.000*		0.788	0.024	0.787	0.016
12-36	2961.7	0.841 *0.000*		**0.842**	0.004	0.840	0.035	0.839	0.034
12-80	2463.02	0.807	0.010	0.806 *0.000*		**0.807**	0.015	0.806	0.009
15-2	3832.04	**0.865** *0.000*		0.864	0.006	0.862	0.006	0.863	0.024
15-3	4272.93	**0.852** *0.000*		0.852	0.013	0.850	0.028	0.850	0.031
15-5	3494.9	**0.842** *0.000*		0.842	0.001	0.840	0.026	0.840	0.013
20-2	8179.02	**0.886**	0.002	0.886 *0.000*		0.883	0.001	0.884	0.008
20-5	13701.8	0.873	0.002	**0.873** *0.000*		0.872	0.003	0.872	0.005
20-6	7182.62	0.894	0.000	**0.894**	0.013	0.892	0.032	0.892	0.032

8 Conclusion

This study considers integer programming based heuristics combining column geneation (CG) and Lagrangian relaxation (LR) with Ant Colony Optimisation (ACO) for a resource constrained job scheduling problem. We see that such a hybrid of decomposition-based methods provide a useful way to provide good feasible solutions to the problem and furthermore provide a guarantee on the quality of these solutions. As such, the results here show that CG or LR combined with ACO can be effective where a problem can be relaxed and there is a reasonable way to repair the solutions.

We find that in a limited amount of CPU time LR-based algorithms are the most effective. However, if the algorithms are allowed to run for a large number of iterations, the CG-based algorithms are preferred. ACO provides assistance with the upper bounds for both CG and LR. It also provides some assistance with the lower bounds in CG. Thus, if the main aim is to provide a solutions with good bounds, CG-ACO is the preferred option whereas if run time is the main constraint, LR-ACO provides a useful way to generate good solutions quickly.

There is still room for improvement in these results. The gaps are often still quite large (smallest gap for 10 or more machines is 12%) suggesting improvements are possible with the lower and upper bounds. The lower bounds are clearly good, however, there is still the possibility to improve them for the largest instances. We are considering introducing additional cuts (e.g. knapsack cuts which disallow jobs from executing at the same time) which may help improve the bounds or convergence. Clearly, the upper bounds can also still be improved and the guidance provided by the improved formulation.

Since these algorithms require large run-times, investigating parallel implementations would be beneficial. There are two obvious ways in which a parallel implementation could work. Firstly, each sub-problem could be run in parallel and secondly, the ants could be constructed in parallel. Solving the sub-problems in parallel are likely to be most effective since the main run-time overheads are incurred here. Thus, a parallel implementation could significantly reduce the run-time in providing the same quality solutions.

References

1. Ballestin, F., Trautmann, N.: An Iterated-local-search Heuristic for the Resource-constrained Weighted Earliness-tardiness Project Scheduling Problem. International Journal of Production Research 46, 6231–6249 (2008)
2. Barnhart, C., Johnson, E.L., Nemhauser, G.L., Savelsbergh, M.W.P., Vance, P.H.: Branch-and-Price: Column Generation for Solving Huge Integer Programs. Operations Research 46(3), 316–329 (1998)
3. Bazaraa, M.S., Jarvis, J.J., Sherali, H.F.: Linear Programming and Network Flows, 2nd edn. John Wiley & Sons, New York (1990)
4. Bertsekas, D.: Nonlinear Programming, 2nd edn. Athena Scientific, New Hampshire (1995)
5. Bertsekas, D.P., Nedić, A., Ozdaglar, A.E.: Convex Analysis and Optimization. Athena Scientific (2003)

6. Blum, C., Roli, A.: Metaheuristics in Combinatorial Optimization: Overview and Conceptual Comparison. ACM Computing Surveys 35, 268–308 (2003)
7. Boschetti, M., Maniezzo, V.: Benders Decomposition, Lagrangean Relaxation and Metaheuristic Design. Journal of Heuristics 15(3), 283–312 (2009)
8. Brucker, P., Drexl, A., Möhring, R., Neumann, K., Pesch, E.: Resource-constrained Project Scheduling: Notation, Classification, Models, and Methods. European Journal of Operational Research 112, 3–41 (1999)
9. Demeulemeester, E., Herroelen, W.: Project Scheduling: A Research Handbook. Kluwer, Boston (2002)
10. Dorigo, M.: Optimization, Learning and Natural Algorithms. Ph.D. thesis, Dip. Elettronica (1992)
11. Dorigo, M., Stützle, T.: Ant Colony Optimization. MIT Press, Cambridge (2004)
12. Fisher, M.: The Lagrangian Relaxation Method for Solving Integer Programming Problems. Management Science 50(12), 1861–1871 (2004)
13. Massen, F., Deville, Y., Van Hentenryck, P.: Pheromone-Based Heuristic Column Generation for Vehicle Routing Problems with Black Box Feasibility. In: Beldiceanu, N., Jussien, N., Pinson, É. (eds.) CPAIOR 2012. LNCS, vol. 7298, pp. 260–274. Springer, Heidelberg (2012)
14. Massen, F., López-Ibáñez, M., Stützle, T., Deville, Y.: Experimental Analysis of Pheromone-Based Heuristic Column Generation Using irace. In: Blesa, M.J., Blum, C., Festa, P., Roli, A., Sampels, M. (eds.) HM 2013. LNCS, vol. 7919, pp. 92–106. Springer, Heidelberg (2013)
15. du Merle, O., Villeneuve, D., Desrosiers, J., Hansen, P.: Stabilized column generation. Discrete Mathematics 194, 229–237 (1997)
16. Neumann, K., Schwindt, C., Zimmermann, J.: Project Scheduling with Time Windows and Scarce Resources. Springer, Berlin (2003)
17. Singh, G., Ernst, A.T.: Resource Constraint Scheduling with a Fractional Shared Resource. Operations Research Letters 39(5), 363–368 (2011)
18. Singh, G., Weiskircher, R.: Collaborative resource constraint scheduling with a fractional shared resource. In: IEEE/WIC/ACM International Conference on Web Intelligence and Intelligent Agent Technology, vol. 2, pp. 359–365. IEEE (2008)
19. Singh, G., Weiskircher, R.: A Multi-Agent System for Decentralised Fractional Shared Resource Constraint Scheduling. Web Intelligence and Agent Systems 9(2), 99–108 (2011)
20. Thiruvady, D., Ernst, A.T., Singh, G.: Parallel Ant Colony Optimization for Resource Constrained Job Scheduling. Annals of Operations Research, 1–18 (2014)
21. Thiruvady, D., Ernst, A.T., Wallace, M.: A Lagrangian-ACO Matheuristic for Car Sequencing (to be published, 2014)
22. Thiruvady, D., Singh, G., Ernst, A.T., Meyer, B.: Constraint-based ACO for a Shared Resource Constrained Scheduling Problem. International Journal of Production Economics 141(1), 230–242 (2012)
23. Thiruvady, D., Wallace, M., Gu, H., Schutt, A.: A Lagrangian Relaxation and ACO Hybrid for Resource Constrained Project Scheduling with Discounted Cash Flows (to be published, 2014)
24. Wolsey, L.A.: Integer programming. Wiley-Interscience, New York (1998)

Iterative Probabilistic Tree Search for the Minimum Common String Partition Problem

Christian Blum[1,2], José A. Lozano[1], and Pedro Pinacho Davidson[1,3]

[1] Department of Computer Science and Artifical Intelligence,
University of the Basque Country UPV/EHU, San Sebastian, Spain
{christian.blum,ja.lozano}@ehu.es
[2] IKERBASQUE, Basque Foundation for Science, Bilbao, Spain
[3] Escuela de Informática, Universidad Santo Tomás, Concepción, Chile
ppinacho@santotomas.cl

Abstract. The minimum common string partition problem is an NP-hard combinatorial optimization problem with applications in computational biology. In this work we propose an iterative probabilistic tree search algorithm for tackling this problem. By means of an extensive experimental evaluation we show the superiority of our approach in comparison to a standard greedy algorithm and a metaheuristic based on ant colony optimization from the related literature.

1 Introduction

Optimization problems related to strings—such as, for example, DNA sequences—are very common in bioinformatics. Examples include the longest common subsequence problem and its variants [13,21], string consensus problems such as the far-from most string problem [19,18], and alignment problems [11]. Many of these problems are computationally very difficult, if not even NP-hard. In this work we deal with the *minimum common string partition* (MCSP) problem. In this problem, we are given two related input strings which must both be partitioned into the same collection of substrings. The size of the collection is subject to minimization. Note that a formal description of the problem will be provided in Section 1.1. Chen et al. [2] point out, for example, that the MCSP problem is closely related to the problem of sorting by reversals with duplicates, a key problem in genome rearrangement.

In this paper we introduce an iterative, probabilistic variant of a known greedy heuristic for the MCSP problem. The construction of solutions in this algorithm is similar to the solution construction phase of a greedy randomized adaptive search procedure (GRASP) [7]. Therefore, the algorithm belongs to the class of metaheuristics. The proposed algorithm also makes use of parallel solution constructions in the search tree defined by the given problem instances and the solution construction mechanism. This means that algorithm combines algorithmic components of metaheuristics with those ones of complete search. Therefore, the algorithm can be seen as a so-called hybrid metaheuristic [1]. The obtained results show that the performance of the proposed algorithm outperforms the current state of the art.

M.J. Blesa, C. Blum, and S. Voß (Eds.): HM 2014, LNCS 8457, pp. 145–154, 2014.

1.1 Problem Description

The MCSP problem can technically be stated as follows. Given are two related input strings, s_1 and s_2, of length n over a finite alphabet Σ. In this context, note that two strings are called *related* if each letter appears the same number of times in each of the two strings. This definition implies that s_1 and s_2 have the same length. A valid solution to the problem is obtained by partitioning s_1 into a set P_1 of non-overlapping substrings, and s_2 into a set P_2 of non-overlapping substrings, such that $P_1 = P_2$. Moreover, we are interested in finding a valid solution such that $|P_1| = |P_2|$ is minimal.

Consider the following example. Given are DNA sequences $s_1 = \mathbf{AGACTG}$ and $s_2 = \mathbf{ACTAGG}$. Obviously, s_1 and s_2 are related because $\mathbf{A}$ and $\mathbf{G}$ appear twice in both input strings, while $\mathbf{C}$ and $\mathbf{T}$ appear once. A trivial valid solution can be obtained by partitioning both strings into substrings of length 1, that is, $P_1 = P_2 = \{\mathbf{A}, \mathbf{A}, \mathbf{G}, \mathbf{G}, \mathbf{C}, \mathbf{T}\}$. The objective function value of this solution is 6. However, the optimal solution, with objective function value 3, is $P_1 = P_2 = \{\mathbf{ACT}, \mathbf{AG}, \mathbf{G}\}$.

1.2 Related Work

The MCSP problem has been introduced by Chen et al. [2] due to its relation to genome rearrangement. More specifically, it has applications in biological questions such as: May a given DNA string possibly be obtained by rearrangements of another DNA string? The general problem has been shown to be NP-hard even in very restrictive cases [9]. Other papers concerning problem hardness consider, for example, the k-MCSP problem, which is the version of the MCSP problem in which each letter occurs at most k times in each input string. The 2-MCSP problem was shown to be APX-hard in [9]. When the input strings are over an alphabet of size c, the corresponding problem is denoted as $MCSP^c$. Jiang et al. proved that the decision version of the $MCSP^c$ problem is NP-complete when $c \geq 2$ [14].

The MCSP has been considered quite extensively by researchers dealing with the approximability of the problem. Cormode and Muthukrishnan [4], for example, proposed an $O(log n log^* n)$-approximation for the *edit distance with moves* problem, which is a more general case of the MCSP problem. Shapira and Storer [20] extended on this result. Other approximation approaches for the MCSP problem have been proposed in [17]. In this context, Chrobak et al. [3] studied a simple greedy approach for the MCSP problem, showing that the approximation ratio concerning the 2-MCSP problem is 3, and for the 4-MCSP problem the approximation ratio is $\Omega(log(n))$. In the case of the general MCSP problem, the approximation ratio is between $\Omega(n^{0.43})$ and $O(n^{0.67})$, assuming that the input strings use an alphabet of size $O(log(n))$. Later Kaplan and Shafir [15] raised the lower bound to $\Omega(n^{0.46})$. Kolman proposed a modified version of the simple greedy algorithm with an approximation ratio of $O(k^2)$ for the k-MCSP [16]. Recently, Goldstein and Lewenstein proposed a greedy algorithm for the MCSP problem that runs in $O(n)$ time (see [10]). He [12] introduced a greedy algorithm with the aim of obtaining better average results.

Damaschke [5] was the first one to study the fixed-parameter tractability (FPT) of the problem. Later, Jiang et al. [14] showed that both the k-MCSP and MCSP^c problems admit FPT algorithms when k and c are constant parameters. Finally, Fu et al. [8] proposed a $O(2^n n^{O(1)})$ time algorithm for the general case and an $O(n(logn)^2)$ time algorithm applicable under some constraints.

To our knowledge, the only metaheuristic algorithm which has been proposed in the related literature for the MCSP problem is the $\mathcal{MAX}$-$\mathcal{MIN}$ Ant System by Ferdous and Sohel [6]. The authors applied their algorithm to a range of artificial and real DNA.

1.3   Organization of the Paper

The remainder of this work is organized as follows. Section 2 provides a description of the proposed algorithm. Furthermore, Section 3 presents a detailed study of the performance of the proposed algorithm in comparison to the state of the art. Finally, in Section 4 we provide conclusions and an outlook to future work.

2   Proposed Algorithm

In the following the proposed algorithm is described. However, before the main algorithm can be described, it is necessary to introduce a number of definitions.

2.1   Preliminaries

Henceforth, a *common block* b of input strings s_1 and s_2 is denoted as a triple (t^b, i^b, j^b) where t^b is a string which can be found starting at position $1 \leq i^b \leq n$ in string s_1 and starting at position $1 \leq j^b \leq n$ in string s_2. Moreover, let B be the set of all possible common blocks of s_1 and s_2. Then, a valid solution $\mathcal{S}$ to the MCSP problem is a subset of B such that:

1. $\sum_{b \in \mathcal{S}} |t^b| = n$, that is, the sum of the length of the common blocks is equal to the length of the input strings
2. For any two common blocks $b, b' \in \mathcal{S}$ it holds that they neither overlap in s_1 nor in s_2

Moreover, a valid partial solution $\mathcal{S}_{\text{partial}}$ is a subset of B such that $\sum_{b \in \mathcal{S}_{\text{partial}}} |t^b| < n$ and for any two common blocks $b, b' \in \mathcal{S}_{\text{partial}}$ it holds that they neither overlap in s_1 nor in s_2. Note that any valid partial solution can be extended to be a valid solution. Furthermore, given a partial solution $\mathcal{S}_{\text{partial}}$, set $B(\mathcal{S}_{\text{partial}}) \subset B$ denotes the set of common blocks that may be used in order to extend $\mathcal{S}_{\text{partial}}$ such that the result is again a valid partial solution. Finally, $B_{\max}(\mathcal{S}_{\text{partial}}) \subseteq B(\mathcal{S}_{\text{partial}})$ is the subset of $B(\mathcal{S}_{\text{partial}})$ which contains the common blocks of maximal size.

2.2   Main Algorithm

The proposed algorithm is sketched in Algorithm 1. Apart from the two input strings s_1 and s_2, the algorithm requires the setting of three paramters: (1) b_{size}, henceforth referred to as *beam size*; (2) d_{rate}, henceforth called the *determinism rate*; and (3) l_{size}, the so-called *candidate list size*.

While the CPU time limit is not reached, the algorithm repeats the following procedure over and over again. The set of current partial solutions, $\mathcal{P}$, is initialized with an empty partial solution. Then, while the size of the current set of partial solutions does not surpass b_{size}, the algorithm iteratively extends all partial solutions from $\mathcal{P}$ deterministically with all common blocks from $B_{\max}(S_{\text{partial}})$. Remember that this set contains all common blocks from $B(S_{\text{partial}})$ of maximal length. Once this stage of the algorithm ends, the algorithm completes each of the partial solutions S_{partial} from $\mathcal{P}$ in the following way. It each step, a random value $\delta \in [0,1]$ is chosen. If this value is smaller or equal than d_{rate}, the determinism rate, a deterministic construction step is performed by selecting the longest common block (b^*) from $B(S_{\text{partial}})$ and adding b^* to S_{partial}. Otherwise, a probabilistic construction step is performed by, first, selecting a subset L of $B(S_{\text{partial}})$ of (at most) size l_{size} that contains the longest common blocks from $B(S_{\text{partial}})$. S_{partial} is then extended by a common block randomly chosen from L. The output of the algorithm consists in the best complete solution that was generated by the algorithm within the allowed CPU time.

3   Experimental Evaluation

TREESEARCH was implemented in ANSI C++ using GCC 4.7.3 for compiling the software. The experimental results that we outline in the following were obtained on a cluster of PCs with "Intel(R) Xeon(R) CPU 5130" CPUs of 4 nuclii of 2000 MHz and 4 Gigabyte of RAM. In the following we first describe the benchmark set that we used for the experimental evaluation. After the description of the algorithm tuning, the numerical results are presented.

3.1   Problem Instances

Each problem instance consists of two related input strings. For testing the algorithm we chose the same set of benchmark instances that was used by Ferdous and Sohel in [6] for the experimental evaluation of their ant colony optimization approach. This set contains, in total, 30 artificial instances and 15 real-life instances consisting of DNA sequences. Moreover, the benchmark set consists of four subsets of instances. The first subset (henceforth labelled GROUP1) consists of 10 artificial instances in which the input strings are maximally of length 200. The second set (GROUP2) consists of 10 artificial instances with input string lengths between 201 and 400. In the third set (GROUP3) the input strings of the 10 artificial instances have lengths between 401 and 600. Finally, the fourth set (REAL) consists of 15 real-life instances of various lengths.

Algorithm 1. Iterative probabilistic treesearch (TREESEARCH) for the MSCP problem

---

1: **input:** s_1, s_2, b_{size}, d_{rate}, l_{size}
2: $S_{\text{bsf}} :=$ NULL
3: **while** CPU time limit not reached **do**
4:　　$S_{\text{partial}} := \emptyset$
5:　　$\mathcal{P} := \{S_{\text{partial}}\}$
6:　　**while** $|\mathcal{P}| < b_{\text{size}}$ **and not** $\mathcal{P} = \emptyset$ **do**
7:　　　　$\mathcal{P}_{\text{new}} := \emptyset$
8:　　　　**for all** $S_{\text{partial}} \in \mathcal{P}$ **do**
9:　　　　　　**for all** $b \in B_{\max}(S_{\text{partial}})$ **do**
10:　　　　　　　$S'_{\text{partial}} := S_{\text{partial}} \cup \{b\}$
11:　　　　　　　**if** S'_{partial} is a complete solution **then**
12:　　　　　　　　**if** $|S'_{\text{partial}}| < |S_{\text{bsf}}|$ **then** $S_{\text{bsf}} := S'_{\text{partial}}$ **end if**
13:　　　　　　　**else**
14:　　　　　　　　$\mathcal{P}_{\text{new}} := \mathcal{P}_{\text{new}} \cup \{S'_{\text{partial}}\}$
15:　　　　　　　**end if**
16:　　　　　　**end for**
17:　　　　**end for**
18:　　　$\mathcal{P} := \mathcal{P}_{\text{new}}$
19:　　**end while**
20:　　**for all** $S_{\text{partial}} \in \mathcal{P}$ **do**
21:　　　　**while** S_{partial} is not a complete solution **do**
22:　　　　　　Choose a random value $\delta \in [0, 1]$
23:　　　　　　**if** $\delta \leq d_{\text{rate}}$ **then**
24:　　　　　　　Choose b^* such that $|t^{b^*}| \geq |t^b|$ for all $b \in B(S_{\text{partial}})$
25:　　　　　　　$S_{\text{partial}} := S_{\text{partial}} \cup \{b^*\}$
26:　　　　　　**else**
27:　　　　　　　Let $L \subseteq B(S_{\text{partial}})$ contain the (at most) l_{size} longest common blocks from $B(S_{\text{partial}})$
28:　　　　　　　Choose randomly b^* from L
29:　　　　　　　$S_{\text{partial}} := S_{\text{partial}} \cup \{b^*\}$
30:　　　　　　**end if**
31:　　　　**end while**
32:　　　　**if** $|S_{\text{partial}}| < |S_{\text{bsf}}|$ **then** $S_{\text{bsf}} := S_{\text{partial}}$ **end if**
33:　　**end for**
34: **end while**
35: **output:** S_{bsf} (the best solution found)

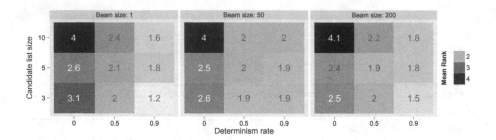

Fig. 1. Graphical presentation of the tuning results. See the text for an analysis of the graphics.

3.2   Algorithm Tuning

The three algorithm parameters—that is, b_{size}, d_{rate} and l_{size}, as described in Section 2.2—were considered for parameter tuning. More specifically, for b_{size} we considered values from $\{1, 50, 200\}$, for d_{rate} values from $\{0.0, 0.5, 0.9\}$, and for l_{size} from $\{3, 5, 10\}$. This results in a total of 27 different parameter value combinations. For each of these 27 settings, the algorithm was applied—with a run time limit of 1000 CPU seconds—exactly once to the first and to the last problem instance of each of the four instance groups.

The following procedure was applied for displaying the results. First, we separated the results into two sets. The first set contains the results of all parameter settings with $b_{\text{size}} = 1$, which is the setting that does not make use of the parallel construction of solutions (lines 6–19 of Algorithm 1). The second set contains the rest of the results. Then, within each set the results are ranked concerning each of the eight considered problem instances. And finally, an average rank is computed for each parameter combination. These average ranks are shown (for both sets of results) in Figure 1. Additionally, the color of a grid cell indicates the quality of the corresponding parameter setting. In general, the darker the color the lower the quality of the corresponding parameter setting. The displayed graphic contains three grids. In each of the grids the x-axis represents the three values for d_{rate} (the determinism rate), while the y-axis represents the three values for l_{size} (the candidate list size). The grid on the left presents the average ranks of all parameter settings with $b_{\text{size}} = 1$, while the other two grids present the average ranks of the parameter settings with $b_{\text{size}} = 50$ (grid in the middle), respectively $b_{\text{size}} = 200$ (grid on the right).

The following conclusions can be drawn from the results. First, the quality of the results seems to improve with a growing determinism rate (d_{rate}). Second, the candidate list size (l_{size}) should be rather small. In summary, the best-ranked parameter setting when $b_{\text{size}} = 1$ is ($d_{\text{rate}} = 0.9, l_{\text{size}} = 3$), and the best-ranked parameter setting when $b_{\text{size}} > 1$ is ($b_{\text{size}} = 200, d_{\text{rate}} = 0.9, l_{\text{size}} = 3$). In fact, these are the parameter values that we selected for the final experimental evaluation. Henceforth we refer to the algorithm using the former set of parameters—that

Table 1. Results for the 10 instances of GROUP1

# of instance	GREEDY value	ACO average	TREESEARCH1 best	average (std.)	time	TREESEARCH2 best	average (std.)	time
1	46	42.75	42	42.50 (0.53)	197.36	42	42.30 (0.48)	285.63
2	56	51.50	48	48.90 (0.32)	175.95	49	49.00 (0.00)	91.96
3	62	56.75	56	56.00 (0.00)	253.47	56	56.00 (0.00)	123.97
4	46	43.00	43	43.00 (0.00)	52.34	43	43.00 (0.00)	88.72
5	44	43.00	41	41.00 (0.00)	124.53	41	41.00 (0.00)	30.48
6	48	42.25	41	41.10 (0.32)	278.28	41	41.70 (0.48)	221.10
7	65	60.00	60	60.80 (0.42)	106.12	61	61.00 (0.00)	21.34
8	51	47.00	45	45.30 (0.48)	389.73	45	45.30 (0.48)	369.13
9	46	45.75	43	43.00 (0.00)	247.70	43	43.00 (0.00)	166.10
10	63	59.25	58	58.80 (0.42)	218.88	59	59.00 (0.00)	76.57

is, the ones for $b_{size} = 1$—with TREESEARCH1, and to the algorithm using the latter set of parameters with TREESEARCH2.

3.3 Numerical Results

Both TREESEARCH1 and TREESEARCH2 were applied with a run time limit of 1000 CPU seconds to all 45 problem instances. 10 repetitions were performed for each instance. The numerical results are presented in Tables 1–4. The layout of all four tables is as follows. The first column provides the instance number. The second column contains the results of the greedy algorithm from [3] (results were taken from [6]). The third column provides the results (averaged over four independent runs) of the $\mathcal{MAX}$-$\mathcal{MIN}$ Ant System (henceforth simply labelled ACO) by Ferdous and Sohel [6]. Concerning ACO, each run was performed with a CPU time limit of 2 hours on a computer with an "Intel(R) 2 Quad" CPU with 2.33 GHz and 4 GB of RAM. Note that in the case of Table 4 we also provide the values of the best solutions found by ACO which were obtained by personal communication with the authors of [6]. Finally, the results of TREESEARCH1 and TREESEARCH2 are presented in three columns for each algorithm version. The first of these four columns provides the value of the best solution found in 10 independent runs. The second column provides the average solution quality obtained together with the corresponding standard deviation (in brackets). The third column gives the average time at which the best solution of a run was found. Finally, note that for each problem instance the result of the best-performing algorithm (concering the average solution quality) is marked with a grey background.

The following conclusions can be drawn from the results. First, both version of TREESEARCH outperform, in general, both GREEDY and ACO. Concerning the average solution quality, TREESEARCH1 outperforms ACO in 39 out of 45 cases, while TREESEARCH2 outperforms ACO in 41 out of 45 cases. Based on the numerical results shown in Tables 1–4 we computed the average percentage improvement of the TREESEARCH versions over ACO, averaged over the four instance groups. These numbers can be seen in Table 5. In all cases, the improvements are between two and three percent.

Table 2. Results for the 10 instances of GROUP2

# of instance	GREEDY value	ACO average	TREESEARCH1 best	average (std.)	time	TREESEARCH2 best	average (std.)	time
1	119	114.25	112	112.80 (0.42)	236.02	111	112.10 (0.74)	270.48
2	122	119.00	114	115.60 (0.70)	471.67	115	115.60 (0.52)	466.58
3	114	112.25	107	108.30 (0.67)	207.00	107	107.60 (0.52)	501.57
4	116	116.25	111	112.40 (0.70)	291.36	111	112.40 (0.70)	206.16
5	135	132.25	127	128.70 (1.16)	373.68	128	129.50 (0.85)	379.25
6	108	105.50	103	103.60 (0.52)	353.94	102	103.20 (0.63)	229.14
7	108	99.00	96	96.90 (0.32)	327.40	96	96.70 (0.67)	318.24
8	123	118.00	115	115.10 (0.32)	369.12	114	115.30 (0.67)	305.54
9	124	119.50	114	114.80 (0.63)	235.29	113	114.50 (0.97)	281.77
10	105	101.75	98	98.60 (0.52)	162.48	98	98.70 (0.48)	308.61

Table 3. Results for the 10 instances of GROUP3

# of instance	GREEDY value	ACO average	TREESEARCH1 best	average (std.)	time	TREESEARCH2 best	average (std.)	time
1	182	180.00	171	172.90 (1.20)	196.92	171	172.90 (0.88)	434.40
2	175	176.25	168	170.80 (1.23)	390.75	170	170.70 (0.48)	396.32
3	196	188.00	185	186.30 (0.67)	361.04	186	186.80 (0.63)	446.02
4	192	184.25	179	181.00 (0.94)	335.72	179	180.50 (0.85)	423.51
5	176	171.75	164	165.00 (0.47)	399.88	163	164.70 (0.82)	437.65
6	170	163.25	163	164.40 (0.70)	427.94	162	164.40 (0.97)	506.95
7	173	168.50	161	162.40 (0.84)	488.96	161	162.60 (0.84)	474.20
8	185	176.25	172	172.40 (0.52)	316.08	169	171.90 (1.37)	376.09
9	174	172.75	170	170.60 (0.52)	365.62	169	170.40 (0.84)	368.80
10	171	167.25	161	162.50 (0.85)	346.42	161	162.30 (0.82)	483.19

Table 4. Results for the 15 instances of set REAL

# of instance	GREEDY value	ACO best	average	TREESEARCH1 best	average (std.)	time	TREESEARCH2 best	average (std.)	time
1	95	87	87.75	87	87.80 (0.42)	314.42	86	87.30 (0.67)	332.09
2	161	155	158.50	154	154.50 (0.53)	384.93	155	155.50 (0.53)	424.61
3	121	116	116.50	113	113.80 (0.63)	268.66	113	113.80 (0.63)	430.52
4	173	164	164.75	159	160.60 (0.84)	360.61	158	160.30 (1.16)	436.76
5	172	171	171.75	166	167.80 (1.03)	521.06	165	167.60 (1.07)	375.17
6	153	145	146.00	144	144.90 (0.80)	212.69	143	144.10 (0.74)	365.66
7	140	140	140.75	131	133.00 (0.94)	425.30	131	132.50 (0.71)	286.02
8	134	130	131.00	128	128.70 (0.48)	414.49	128	128.90 (0.57)	482.46
9	149	146	148.50	142	142.60 (0.52)	314.78	142	142.70 (0.67)	330.21
10	151	148	149.00	144	145.30 (0.82)	465.11	145	145.60 (0.52)	274.35
11	126	124	124.50	121	121.60 (0.52)	464.24	121	121.70 (0.48)	331.92
12	143	137	138.25	138	139.00 (0.82)	360.15	138	139.40 (0.70)	256.56
13	180	180	181.00	171	173.20 (1.14)	417.10	172	173.20 (0.92)	455.36
14	152	147	147.75	147	147.80 (0.63)	367.72	146	147.30 (0.67)	465.47
15	157	160	161.25	152	153.20 (0.92)	313.26	152	153.10 (0.57)	389.08

Concerning a comparison between TREESEARCH1 and TREESEARCH2, no significant difference can be observed. However, it is clearly the case that some problem instances are more easily solved by TREESEARCH1—see, for example, instance 5 of GROUP2 or instance 2 of REAL—while other problem instances should better be solved by TREESEARCH2, such as, for example, instance 1 of GROUP2 or instance 6 of REAL. However, the results do not permit to make general claims in this context.

Table 5. Average percentage improvement of the TREESEARCH variants over ACO per instance group

Algorithm version	GROUP1	GROUP2	GROUP3	REAL
TREESEARCH1	2.34%	2.71%	2.28%	2.34%
TREESEARCH2	2.16%	2.82%	2.35%	2.39%

4   Conclusions and Future Work

In this work we introduced an iterative probabilistic tree search algorithm for the so-called minimum common string partition problem. The tackled problem, which has applications in computational biology, is an NP-hard combinatorial optimization problem. The proposed algorithm can be seen as a hybrid technique, due to the fact that it combines elements from metaheuristic search methods with algorithmic components originating from complete search. Two different versions of the algorithm were tested. In comparison with a recently proposed ant colony optimization approach from the literature, our algorithms have proved to be very competitive, outperforming the ant colony optimization approach in 39 (respectively 41) out of 45 cases.

Future work will deal with the use of bounding information within tree search for obtaining a better guidance of the search process. We also plan to invest some time into the development of complete methods in order to generate optimal solutions.

Acknowledgments. C. Blum was supported by project TIN2012-37930 of the Spanish Government. In addition, support is acknowledged from IKERBASQUE (Basque Foundation for Science). J. A. Lozano was partially supported by the Saiotek and IT609-13 programs (Basque Government), TIN2010-14931 (Spanish Ministry of Science and Innovation), COMBIOMED network in computational bio-medicine (Carlos III Health Institute).

References

1. Blum, C., Puchinger, J., Raidl, G., Roli, A.: Hybrid metaheuristics in combinatorial optimization: A survey. Applied Soft Computing 11(6), 4135–4151 (2011)
2. Chen, X., Zheng, J., Fu, Z., Nan, P., Zhong, Y., Lonardi, S., Jiang, T.: Computing the assignment of orthologous genes via genome rearrangement. In: Proceedings of the Asia Pacific Bioinformatics Conference 2005, pp. 363–378 (2005)
3. Chrobak, M., Kolman, P., Sgall, J.: The greedy algorithm for the minimum common string partition problem. In: Jansen, K., Khanna, S., Rolim, J.D.P., Ron, D. (eds.) RANDOM 2004 and APPROX 2004. LNCS, vol. 3122, pp. 84–95. Springer, Heidelberg (2004)
4. Cormode, G., Muthukrishnan, S.: The string edit distance matching problem with moves. ACM Transactions on Algorithms 3(2), 1–19 (2007)

5. Damaschke, P.: Minimum common string partition parameterized. In: Crandall, K.A., Lagergren, J. (eds.) WABI 2008. LNCS (LNBI), vol. 5251, pp. 87–98. Springer, Heidelberg (2008)
6. Ferdous, S.M., Rahman, M.S.: Solving the minimum common string partition problem with the help of ants. In: Tan, Y., Shi, Y., Mo, H. (eds.) ICSI 2013, Part I. LNCS, vol. 7928, pp. 306–313. Springer, Heidelberg (2013)
7. Festa, P., Resende, M.: GRASP: An annotated bibliography. In: Ribeiro, C., Hansen, P. (eds.) Essays and Surveys on Metaheuristics, pp. 325–367. Kluwer Academic Publishers (2002)
8. Fu, B., Jiang, H., Yang, B., Zhu, B.: Exponential and polynomial time algorithms for the minimum common string partition problem. In: Wang, W., Zhu, X., Du, D.-Z. (eds.) COCOA 2011. LNCS, vol. 6831, pp. 299–310. Springer, Heidelberg (2011)
9. Goldstein, A., Kolman, P., Zheng, J.: Minimum common string partition problem: Hardness and approximations. In: Fleischer, R., Trippen, G. (eds.) ISAAC 2004. LNCS, vol. 3341, pp. 484–495. Springer, Heidelberg (2004)
10. Goldstein, I., Lewenstein, M.: Quick greedy computation for minimum common string partitions. In: Giancarlo, R., Manzini, G. (eds.) CPM 2011. LNCS, vol. 6661, pp. 273–284. Springer, Heidelberg (2011)
11. Gusfield, D.: Algorithms on Strings, Trees, and Sequences. Computer Science and Computational Biology. Cambridge University Press, Cambridge (1997)
12. He, D.: A novel greedy algorithm for the minimum common string partition problem. In: Măndoiu, I.I., Zelikovsky, A. (eds.) ISBRA 2007. LNCS (LNBI), vol. 4463, pp. 441–452. Springer, Heidelberg (2007)
13. Hsu, W.J., Du, M.W.: Computing a longest common subsequence for a set of strings. BIT Numerical Mathematics 24(1), 45–59 (1984)
14. Jiang, H., Zhu, B., Zhu, D., Zhu, H.: Minimum common string partition revisited. Journal of Combinatorial Optimization 23(4), 519–527 (2012)
15. Kaplan, H., Shafrir, N.: The greedy algorithm for edit distance with moves. Information Processing Letters 97(1), 23–27 (2006)
16. Kolman, P.: Approximating reversal distance for strings with bounded number of duplicates. In: Jedrzejowicz, J., Szepietowski, A. (eds.) MFCS 2005. LNCS, vol. 3618, pp. 580–590. Springer, Heidelberg (2005)
17. Kolman, P., Waleń, T.: Reversal distance for strings with duplicates: Linear time approximation using hitting set. In: Erlebach, T., Kaklamanis, C. (eds.) WAOA 2006. LNCS, vol. 4368, pp. 279–289. Springer, Heidelberg (2007)
18. Meneses, C., Oliveira, C., Pardalos, P.: Optimization techniques for string selection and comparison problems in genomics. IEEE Engineering in Medicine and Biology Magazine 24(3), 81–87 (2005)
19. Mousavi, S., Babaie, M., Montazerian, M.: An improved heuristic for the far from most strings problem. Journal of Heuristics 18, 239–262 (2012)
20. Shapira, D., Storer, J.A.: Edit distance with move operations. In: Apostolico, A., Takeda, M. (eds.) CPM 2002. LNCS, vol. 2373, pp. 85–98. Springer, Heidelberg (2002)
21. Smith, T., Waterman, M.: Identification of common molecular subsequences. Journal of Molecular Biology 147(1), 195–197 (1981)

JAM: A Tabu-Based Two-Stage Simulated Annealing Algorithm for the Multidimensional Arrangement Problem*

Jordi Arjona Aroca[1] and Antonio Fernández Anta[2]

[1] Universidad Carlos III de Madrid, Madrid, Spain
[2] IMDEA Networks Institute, Madrid, Spain

Abstract. In this paper we study a version of the Multidimensional Arrangement Problem (MAP) that embeds a graph into a multidimensional array minimizing the aggregated (Manhattan) distance of the embedded edges. This problem includes the minimum Linear Arrangement Problem (minLA) as a special case, among others. We propose JAM, a tabu-based two-stage simulated annealing heuristic for this problem. Our algorithm relies on existing techniques for the minimum linear arrangement (minLA) problem, which are non-trivially adapted to work in multiple dimensions. Due to the scarcity of specific benchmarks for MAP, we have tested the performance of our algorithm with benchmarks for the minLA and Quadratic Assignment Problems (with more than 80 graphs). For each graph in these benchmarks, we provide results for 1, 2 and 3-dimensional instances of MAP, enlarging, hence, the benchmarking resources for the research community. The results obtained show the practicality of JAM, often matching the best known result and even improving some of them.

Keywords: Multidimensional Arrangement Problem, Minimum Linear Arrangement Problem, Quadratic Assignment Problem, Simulated Annealing, Tabu Search.

1 Introduction

Assignment and arrangement problems have been extensively studied for decades. The most classical and well known application of these problems is the assignment of n facilities to m locations in order to minimize or maximize a certain magnitude, such as cost, flow, etc. In this work, we deal with one of these arrangement problems, the Multidimensional Arrangement Problem (MAP), which was firstly studied by Hansen [12]. MAP covers a great number of applications, such as graph drawing or job scheduling (in 1 dimension), the backboard wiring problem or the arrangement of electronic components in printed circuits (in 2 dimensions), and placing servers in the racks of a data center (in 3 dimensions).

* This research was supported in part by the Comunidad de Madrid grant S2009TIC-1692, Spanish MICINN grant TEC2011-29688-C02-01, and National Natural Science Foundation of China grant 61020106002.

M.J. Blesa, C. Blum, and S. Voß (Eds.): HM 2014, LNCS 8457, pp. 155–168, 2014.

In this paper we focus on the MAP problem, which embeds a graph into a multidimensional array minimizing the aggregated Manhattan distance. To solve this problem, we propose a hybrid simulated annealing heuristic, non-trivially adapting techniques used for the Minimum Linear Arrangement Problem (minLA, the one-dimensional version of MAP) to work in multiple dimensions.

1.1 Problem Definition

Given a graph $G = (V, E)$ and a host D-dimensional array $H(V', E')$ such that $|V'| \geq |V|$, we can define the *Multidimensional Arrangement Problem* as the embedding of G into H, i.e., a mapping of the edges of G to paths in H, such that the aggregated length of the paths in H is minimized. As we will usually work with weighted graphs, the goal is to minimize the weighted sum of the path lengths. Formally, the cost of an embedding $\varphi : V \to V'$ is defined as

$$C(\varphi) = \sum_{(u,v) \in E} w_{uv} \cdot dist\left(\varphi(u), \varphi(v)\right), \tag{1}$$

where w_{uv} is the weight of edge (u, v) and $dist\left(\varphi(u), \varphi(v)\right)$ is the Manhattan distance (the path length) between the images of u and v in the host graph H.

The particular case of $D = 1$ is a well known problem, called the *minimum linear arrangement (minLA)*. In this problem, the objective is to embed a graph onto a one dimensional array. As minLA is known to be NP-complete and MAP has minLA as a special case, it can be concluded that MAP is NP-hard.

1.2 Related Work

The *Quadratic Assignment Problem,* which is a more general problem than MAP, is an NP-hard problem [31] which has been creating interest during more than 50 years [16]. The QAP objective function can be mathematically formulated as follows

$$\sum_{i=1}^{n} \sum_{j=1}^{n} f_{ij} \cdot dist(\pi(i)\pi(j)) + \sum_{i,\pi(i)} b(i, \pi(i)),$$

where f_{ij} is the flow between facilities i and j, $\pi(\cdot)$ is the location at which a facility has been assigned, $dist(x, y)$ denotes the distance between two locations x and y, and $b(i, x)$ is the initial allocation cost of facility i to location x. Many well-known problems, like the traveling salesman problem (TSP), minLA, and MAP, are special cases of QAP.

Some exact algorithms have been developed to solve the QAP problem. However, they are only capable to solve small instances due to the enormous computation capacity required. The largest instances solved optimally surpassed just recently the 100 locations frontier [9], but most of the latest works still work with instances of 30-40 locations [9][26]. These algorithms typically use branch and bound, branch and cut, or dynamic programming.

Approximate methods have also been developed to tackle the QAP problem. We classify them in heuristics and metaheuristics. Starting with heuristics, most of the ones that have been developed can be grouped in constructive, enumeration, and improvement methods. We can find some examples of heuristics applied to the QAP problem in [20,25,11].

Despite of the richness in heuristics, metaheuristics have been attracting most of the attention lately. Most of the metaheuristics applied to the QAP problem can be included in one of the following families: genetic algorithms (GA) [22,8], simulated annealing (SA) [4,38], ant colony optimization (ACO) [32], tabu search (TS) [23,24,33,13], breakout local search (BLS) [2], greedy randomized adaptive search procedures (GRASP) [17], variable neighborhood search (VNS) [39], or hybrid combinations of them [10,34]. Given that QAP is more general than MAP, it is possible to adapt many of these techniques to obtain solutions also for MAP.

Simulated annealing (SA) is a local search based metaheuristic, introduced by Kirkpatrick et al. [15] in 1983. It was inspired in the metallurgical process of annealing, and used to solve combinatorial optimization problems. An SA algorithm is usually described by the following elements: initial solution, neighborhood function, cooling rate, number of iterations per temperature, and stop criteria or final temperature. In a nutshell, SA applied to MAP starts from an initial solution φ_0; and then, in each iteration, a candidate neighboring solution φ_l is chosen, based on a cost-based neighborhood function. Once φ_l is chosen it is compared against the current solution (φ^*) and, depending on whether $\delta = C(\varphi_l) - C(\varphi^*)$ is larger than 0 or not, φ_l is accepted as the new current solution φ^* or tested with an acceptance function. This function depends on the current temperature and is based on the Metropolis criterion [19], that will finally accept φ_l as the new φ^* or refuse it. If a new solution is chosen and it is better than the best-so-far solution φ_{best}, it becomes the new φ_{best}. After running a given number of iterations the system's temperature is cooled down. This process follows until a total number of iterations is run or a termination criteria is met.

Observe that the acceptance function allows the heuristic to admit solutions which are worse than the previous ones. This is generally known as climbing up and helps to avoid that heuristics are trapped in a local optimum. Although the mechanics of SA are not complicated, choosing the cooling rate, stop criteria, and neighborhood function is not trivial.

Simulated annealing was one of the first techniques applied to the QAP problem (c.f., Burkard et al. [4], Wilhelm et al. [38]). We now describe some of the main characteristics of some of the latest works using SA, alone or combined with other techniques. We start with the work of Wang [36], who proposed in 2007 a tabu-based simulated annealing algorithm. In that work, a pure SA algorithm was compared to a tabu-search SA, trying different tabu list sizes and also trying different guided restart and reannealing strategies, enhancing the ability to escape from local optima. In 2012, Wang [37] presented a new work based also on simulated annealing, but trying different guided restart strategies. In both works a local-search-based neighborhood function was used jointly with a geometrical cooling rate schedule (like Kirkpatrick et al. [15]), reheating the algorithm when a restart takes place. In 2012, Jingwei et al. [14] presented a new hybrid algorithm combining ant colonies and simulating annealing. Here, simulated annealing was used to select the best ants in each iteration, while the cooling schedule was also geometrical. In 2003, Misevičius [21] presented a very detailed work comparing multiple previously proposed cooling schedules. With this, he proposed an SA heuristic using a normal-local-search-based neighborhood function, an inhomogeneus annealing cooling schedule without equilibrium tests, like the one proposed by Connolly et al. in [7], and modified reannealing so the cooling schedule oscillates depending on the behavior of the annealing. This heuristic was completed by a post optimization stage based on Taillard's robust tabu search. This heuristic was even able to improve one of the QAPLIB [3] instances.

Finally, Tello et al. [28], in 2008, presented a 2-stage simulated annealing algorithm for the minLA problem. This work was able to improve multiple results from the typical set of minLA benchmarks compiled by Petit [27]. Its main contribution is to design a 2-stage SA algorithm, where the first stage obtains an initial approximation through a frontal increase minimization algorithm, and the second stage is devoted to improve this initial solution. They consider a modified median-based neighborhood function in which the typical 2-exchange strategy is conditioned by the nodes connected to a candidate-to-be-moved node. They also consider different ways of establishing the initial temperature, based on [35], and a different cooling schedule [1]. We will detail these aspects when describing our algorithm in Section 2, as we adopted and adapted some of their ideas for our MAP heuristic.

1.3   Contributions

In this work we present JAM, a tabu-based two-stage simulated annealing heuristic for MAP. In JAM, we use a novel median-based neighborhood function and we non-trivially adapt multiple techniques from the minLA literature to work for multiple dimensions.

Due to the lack of benchmarks specific for the MAP problem, we test our heuristic against minLA and QAP benchmarks, with weighted and unweighted graphs. The minLA benchmark has a one-dimensional array as host graph, while the QAP benchmarks have a 2-dimensional array as host graph. JAM obtains the optimal or best known result in most of the problem instances. Although the benchmarks used were originally for 1 or 2 dimensions, we present results for them for 1, 2 and 3 dimensions, broadening hence the available benchmarks for minLA and QAP as well as creating a benchmark set for MAP. We also present 2 different results for 2 dimensions. The first one is restricted to the case in which guest and host graphs have the same size, i.e., where $|V'| = |V|$. The second one is for a deployment which is more compact (square) and that allows having extra locations, i.e., where $|V'| \geq |V|$.

RoadMap. The paper is organized as follows. In Section 2 we present our own algorithm as well as a detailed description of its different elements. In Section 3 we present the numerical results obtained with our heuristic for multiple benchmarks from the minLA and QAP literature and for which we provide results in 1, 2 and 3 dimensions. We close the paper by presenting some conclusions in Section 4.

2 The Algorithm JAM

In this section we present JAM, a tabu based two-stage simulated annealing algorithm applied to MAP. First, we introduce the notation used throughout the rest of the paper, then provide an overview of JAM, and finally describe each one of its elements.

2.1 Notation

Given a graph $G = (V, E)$, V and E denote its sets of vertices and edges, respectively. Each edge $(i, j) \in E$ has an associated weight denoted by w_{ij}. We denote by A^u the set of nodes adjacent to node u in G.

The host graph $H(V', E')$ is a D-dimensional array. Recall that $|V'| \geq |V|$. The nodes of H are called *locations*. Each location $l \in V'$ is defined by a D-vector $(l_1, l_2, \ldots, l_D)$, where l_i is the dimension i coordinate of location l. On the other hand, $d_i(X)$ is used to denote the dimension i coordinate of all elements of a set X of locations.

In order to improve the cost $C(\varphi)$ of an arrangement φ, JAM performs node "movements." By movement we refer to the action of "moving" node u from a location l to a location l', and "moving" the node v, if any, which is at location l' to l. Formally, this means transforming the arrangement φ into a new φ' such that $\varphi'(x) = \varphi(x)$ for all $x \in V \setminus \{u, v\}$, $\varphi'(u) = l'$, and $\varphi'(v) = l$. There is only a set of valid locations to which a node u can move (see Section 2.3 below), this set is called its *neighborhood* and is denoted N^u.

2.2 Overview of JAM

A sketch of JAM is provided as pseudo code in Algorithm 1. JAM is a two-stage heuristic whose first stage provides an initial solution based on the McAllister heuristic [18]. This arrangement is also the initial best known solution. The current solution and the best known solution are stored in φ^* and φ_{best}, respectively (Lines $1 - 2$). Then, the initial temperature $T(0)$ (Line 3) for the SA is computed. After that, the cooling down process starts, which will take place until the termination criteria are met (Line 4). For every temperature $T(k)$, JAM runs a predefined number of iterations (Line 5), starting with $T(0)$. In each iteration, a node u to be moved is chosen uniformly at random (Line 6). Then, with probability p_N, the set of locations L to which u may be moved is chosen to be the neighborhood N^u (Line 7). Otherwise the only location that will be considered

---

Algorithm 1. JAM Pseudo Code

```
1  φ* ← SetInitialSolution();
2  φ_best ← φ*;
3  k ← 0; T(k) ← SetInitialTemperature();
4  while the termination criteria do not hold do
5  │   for the predefined number of iterations at temperature T(k) do
6  │   │   Choose a node u uniformly at random;
7  │   │   with probability p_N: set L ← N^u;
8  │   │   else set L ← {l}, where l is a location chosen uniformly at random
   │   │   Discard all locations l ∈ L that would lead to a move in the tabu list;
9  │   │   Φ ← {φ_l : φ_l is the arrangement after moving u to l in φ*, ∀l ∈ L};
10 │   │   φ' ← arg min_{φ∈Φ}{C(φ)};
11 │   │   if C(φ') > C(φ*) then
12 │   │   │   φ' ← φ chosen from Φ with probability proportional to C(φ);
13 │   │   δ ← C(φ') − C(φ*);
14 │   │   with probability e^{−δ/T(k)}: φ* ← φ'; φ_best ← arg min{C(φ_best), C(φ*)};
15 │   k ← k + 1; T(k) ← UpdateTemperature(T(k − 1));
16 │   GuidedRestart(φ*, φ_best);
```

---

is a randomly chosen one l (Line 8). Not explicitly shown in Algorithm 1, JAM maintains a tabu list of movements that are not to be redone. Hence, all elements in L that lead to a move in this tabu list are discarded (Line 9). Now, the arrangements Φ resulting of moving u to the remaining locations in L are obtained (Line 10), and the arrangement φ' with the lowest cost among them is chosen (Line 11). If the cost of this φ' is larger than the cost of the current solution φ^*, φ' is replaced by an arrangement chosen from Φ with probabilities proportional to their respective costs (Line $12 - 13$). To complete the iteration, the proposed φ' is adopted with probability $e^{-\delta/T(k)}$, which implies updating φ^* and, if corresponds, φ_{best} (Line 15). (Observe that if $\delta < 0$ then φ' is always adopted.) Once the given number of iterations for $T(k)$ is reached, it is updated (Line 16) and it is decided whether resetting φ^* to φ_{best} is needed (Line 17).

2.3   Elements of JAM

We now provide a detailed description of the elements mentioned above that conform JAM.

First Stage: Initial Solution. McAllister heuristic [18] has been adapted to multiple dimensions and used to obtain an initial solution. McAllister's is a greedy heuristic based on a frontal increase minimization strategy. It chooses a starting node at random and maps it to some location. Then, it greedily maps the rest of nodes. To do so, it maintains three sets of nodes U (Unplaced), P (Placed) and F (Front, the set of placed nodes with at least one neighbor in set U). The next node to be mapped is the one with the least neighbors in set $U \setminus F$, so the front set is minimized.

We implement McAllister's heuristic with all the proposed refinements: a tie breaking strategy, improved initial node selection and deferred node placement (We refer the reader to [18] for further details). Then, we devised a multidimensional allocation technique which maps the first node to location $(0, 0, \ldots, 0)$ in the host graph, and then greedily decides which of the neighboring locations to those of the already allocated nodes is the best position, in terms of cost, for the next node.

Initial Temperature. We decided to initialize the temperature using the same method as Tello et al. [28], which employs the technique proposed by Varanelli and Cohoon [35]. This method approximates the simulated annealing temperature $T(k)$ at which a solution φ^* with cost $C(\varphi^*)$ can be found as best solution. Hence the initial temperature is given by[1]

$$T(0) \approx \left| \frac{\sigma_\infty^2}{C_\infty - C(\varphi^*) - \gamma_\infty \sigma_\infty} \right|,$$

where C_∞ and σ_∞ represent the expected cost and average deviation of the cost over the solution space; $C(\varphi^*)$ represents the cost of the initial solution and γ_∞ represents the difference between the expected cost C_∞ and the best known solution φ_{best} at temperature $T(k)$. γ_∞ can be calculated probabilistically from the number of iterations predefined at each temperature. We refer the reader to [35] for further details.

Cooling Schedule. Our cooling schedule is based on the work from Aarts and Korst [1]. They propose a statistical cooling schedule which depends on the previous temperature, the average deviation of the solutions obtained with the previous temperature $\sigma_{T(k-1)}$, and a tuning parameter λ (such that for small values of λ we obtain small temperature reductions). The cooling schedule is given by the following equation:

$$T(k) = T(k-1) \left(1 + \frac{\log(1 + \lambda)T(k-1)}{3\sigma_{T(k-1)}} \right)^{-1}.$$

Neighboring Solutions. In order to reduce the search space of locations to which a certain node u can be moved, we define a median-based neighborhood function. This function returns a set of neighbors N^u, which is the set of contiguous locations that will be considered for the movement of u. Intuitively, we choose the set N^u to be the locations that minimize the cost of the edges incident in u assuming that only u changes its location (in this fictitious arrangement u may share location with other nodes).

We describe now the process we use to obtain N^u. Let us assume that the nodes adjacent to u in G are $A^u = \{v^1, v^2, \ldots, v^n\}$, and that their respective current location is $l^j = \varphi^*(v^j), \forall j \in [1, n]$. Let us fix one dimension $i \in [1, D]$,

---

[1] We use the absolute value of the expression, unlike in [35], to deal with some cases that resulted in negative values.

and let us sort the nodes in A^u by the dimension i coordinate of their current location, so that $l_i^{j_1} \leq l_i^{j_2} \leq \cdots \leq l_i^{j_n}$. Then, we compute the smallest $m \in [1, n]$ that satisfies

$$\Delta(m) = \sum_{k=1}^{m} w_{uv^{j_k}} - \sum_{k=m+1}^{n} w_{uv^{j_k}} \geq 0.$$

If $\Delta(m) = 0$ then we define a range of values $r_i = [l^{j_m}, l^{j_{m+1}})$. Otherwise, if $\Delta(m) > 0$ then we define the range as the singleton value $r_i = [l^{j_m}]$.

After applying this method to each dimension separately we have ranges $r_1, r_2, \ldots, r_D$. The D-polytope obtained by the combination of these ranges is the set of locations in N^u. I.e., all locations l such that $l_i \in r_i, \forall i \in [1, D]$ belong to N^u. In our implementation of JAM we extended N^u with all the locations that are within distance 2 of the set described, just to increase the movement options.

Evaluating Solutions. We defined the cost of an arrangement $C(\varphi)$ in Eq. 1. However, two different solutions might have the same cost. To consider these cases, we use instead a cost function $C'(\varphi)$ introduced in [29]. The authors there proposed a refined method for estimating the cost of solutions in a minLA problem which considers not only the cost derived from the paths in H but also how the costs of these paths are distributed. The cost of an arrangement φ is then given by

$$C'(\varphi) = \sum_{k=1}^{\Theta} \left(k + \frac{n!}{(n+k)!} \right) e_k, \tag{2}$$

where $\Theta = \sum_{i=1}^{D} d_i - 1$ is the diameter of H and e_k is the number of paths of length k in H. Note that the second term of this formula is always smaller than 1. Then, for solutions where the cost would be the same if we had only considered the first term, the total cost will be smaller if the arrangement has longer paths. A solution with a larger number of longer paths is preferred as it would be, in principle, easier to improve.

Tabu Search (TS). In order to favor the exploration abilities of JAM we incorporate TS. As we said, moving a node u from position l to position l' implies moving the node v in position l', if any, to position l. Our TS mechanism will check that neither u nor v have been in locations l or l', respectively, during the last T_s moves, being T_s the size of our tabu list.

To control when a move is tabu we use a $|V| \times |V'|$ matrix. Every time a node is moved to a certain location, we store the iteration number at which that move was done. If a proposed move has been done during the last T_s iterations, it is discarded. There is one exception to this rule, the *aspiration criterion*. We implemented the most common one: a move will be accepted, despite of being tabu, when it leads to a smaller $C'(\varphi_{best})$.

Guided Restarts. We implement guided restarts in order to help the algorithm to escape from some strong local minima. A restart consists in resetting φ^* to φ_{best}. We decide if a restart is needed after finishing all the iterations at a certain temperature. A restart occurs with probability

$$P(restart) = 1 - e^{\left(-\frac{|\varphi^* - \varphi_{best}|}{\varphi_{best}} \frac{T(0)}{T(k)} \gamma\right)},$$

where $T(0)$ and $T(k)$ are the initial and current temperatures, and γ is a tuning parameter that depends on the size of the graph.

Termination Criteria. We use two termination criteria. Our algorithm will stop when (1) $T(k)$ goes below a predefined temperature threshold T_{th} or when (2) the percentage of accepted moves improving φ^* while at temperature $T(k)$ goes below a second predefined threshold P_{th}. The values for these thresholds depend on the size of the graph.

3   Evaluation of JAM

In this section we present the results obtained for a set of benchmark instances ran in order to evaluate JAM's performance. Ideally, we would have used a set of instances for which we had results in multiple dimensions. However, due to non-existence, to the best of our knowledge, of such a set of instances, we used graphs belonging to benchmarks from the minLA and QAP literature. Our intention, however, is two-fold. First, we want to create such a collection of instances so they can be used in future MAP works as benchmark. Second, by running these graphs in 1 and 2 dimensions, we are broadening the available number of instances and results for both minLA and QAP benchmark collections.

JAM results for graphs from both collections of instances are presented in Tables 2 and 3. We provide two different results for 2 dimensions. First, with either $|V'| = |V|$ or the original configuration (for the case of QAPLIB instances). Second, for a more compact layout allowing that $|V'| \geq |V|$. The *BKV* (Best Known Value) and a δ (the difference between JAM's result and the BKV in percentage) are provided when a BKV is available. In particular, they are given for 1 dimension results in Table 2 and for the first results in 2 dimensions in Table 3. For the 3-dimensional host graphs we chose the number of nodes in each dimension so that the number of empty locations is minimized. In these tables R, C and D denote the number of nodes in the respective dimensions of H (the letters come from rows, columns and depth).

Table 1. Parameters used depending on the number of edges of the graph

Parameter	# Edges				
	≤ 500	≤ 1000	≤ 5000	≤ 10000	≥ 10000
Iterations per Temperature	$2 \cdot 10^5$	$2 \cdot 10^5$	$2.5 \cdot 10^5$	$3 \cdot 10^5$	$3.5 \cdot 10^5$
T_{th}	0.25	0.5	0.75		
P_{th}	0.125	0.075		0.05	
Restart Factor γ	0.3	0.5	1	1.5	

Table 2. Results for the minLA benchmark

Graph	$|V|$	$|E|$	1D			2D			3D			
			BKV	δ	Cost	R	C	Cost	R	C	D	Cost
bcspwr01	39	46	**106**	0%	106	3	13	57	5	8	51	2	4	5	50
bcspwr02	49	59	**161**	0%	161	7	7	72	-		2	5	5	66
bcspwr03	118	179	[588, 679]	-2,50%	662	2	59	384	10	12	255	4	5	6	225
bcsstk01	48	176	**1132**	1,77%	1152	6	8	384	-		2	4	6	314
bcsstk02	66	2145	47916	-0,02%	47905	6	11	12155	8	9	11505	2	3	11	10945
bcsstk04	132	1758	[27569, 29804]	0,03%	29812	11	12	7126	-		3	4	11	5192
bcsstk05	153	1135	[9653, 11057]	0,02%	11059	9	17	11060	11	14	3508	2	7	11	2895
bcsstk22	110	254	-	-	981	10	11	374	-		2	5	11	353
bintree10	1023	1022	3696	0%	3696	31	33	1231	32	32	1233	3	11	31	1098
c1y	828	1749	62230	0.33%	62436	23	36	5760	27	31	5752	6	6	23	3962
can___144	144	576	[2304, 3224]	0%	3224	12	12	1058	-		4	6	6	990
can___161	161	608	[5657, 6696]	0%	6696	7	23	1371	12	14	1253	4	6	7	1012
can___24	24	68	**210**	0%	210	4	6	98	-		2	3	4	98
can___61	61	248	**1137**	0%	1137	1	61	1137	7	9	485	3	3	7	425
can___62	62	78	[187, 212]	-0,94%	210	2	31	130	7	9	90	3	3	7	84
can___73	73	152	[971, 1100]	0%	1100	1	73	1100	7	11	284	3	5	5	229
can___96	96	336	[2105, 2702]	0%	2702	8	12	600	9	11	599	4	4	6	525
curtis54	54	124	**454**	0%	454	6	9	194	7	8	191	3	3	6	179
dwt___162	162	510	[2032, 2431]	0,25%	2437	9	18	812	12	14	814	3	6	9	766
dwt___209	209	767	[5905, 6387]	20,78%	7714	11	19	1588	14	15	1653	5	6	7	1258
dwt___221	221	704	[3603, 3779]	-0,13%	3774	13	17	1184	15	15	1176	5	5	9	1062
dwt___245	245	608	[3422, 3860]	4,53%	4035	7	35	1143	15	17	1054	5	7	7	920
dwt___59	59	104	**289**	0%	289	1	59	289	7	9	134	3	4	5	128
dwt___66	66	127	**192**	0%	192	6	11	164	8	9	163	2	3	11	159
dwt___72	72	75	**167**	0%	167	8	9	78	-		3	4	6	80
dwt___87	87	227	**932**	0%	932	3	29	448	8	11	384	2	4	11	334
fidap005	27	126	**414**	0%	414	5	6	250	-		3	3	3	242
fidapm05			**1003**	0%	1003	6	7	545	-		2	3	7	487
gd95c	62	144	**506**	0%	506	2	31	318	7	9	233	3	3	7	210
gd96b	111	193	1416	0%	1416	3	37	602	10	12	461	4	4	7	380
gd96c	65	125	**519**	0%	519	5	13	196	8	9	188	2	3	11	166
gd96d	180	228	2391	0%	2391	12	15	518	13	14	517	5	6	6	382
ibm32	32	90	**485**	0%	485	4	8	192	5	7	183	2	4	4	155
impcol_b	59	281	[1810, 2076]	0%	2076	1	59	2076	7	9	713	3	4	5	588
lunda	147	1151	[10772, 11323]	0,03%	11326	7	21	2866	11	14	2802	3	7	7	2483
lundb	147	1147	[10712, 11187]	0,04%	11192	7	21	2836	11	14	2787	3	7	7	2452
mesh33x33	1089	2112	31729	3,03%	32693	33	33	2112	-		9	11	11	2764
nos4	100	247	**1031**	0%	1031	10	10	424	-		4	5	5	367
pores_1	30	103	**383**	0%	383	5	6	167	-		2	3	5	147
RandomA1	1000	4974	866968	3,00%	892986	25	40	57855	29	35	57436	10	10	10	26757
RandomA2	1000	24738	6522206	0,44%	6550805	25	40	427480	29	35	415460	10	10	10	196135
steam3	80	424	**1416**	0%	1416	8	10	946	-		4	4	5	842
tub100	100	148	**246**	0%	246	10	10	158	-		4	5	5	152
will57	57	127	**335**	0%	335	3	19	218	7	9	187	2	5	6	180

We provide results for 81 different graphs. We ran JAM a minimum of 5 times per instance, for the sake of statistical significance. We used different configurations that depended on the number of edges of the graph. In particular, the parameters being changed were the predefined number of iterations per temperature, T_{th}, P_{th} and γ. The values used can be found in Table 1. Other parameters used during the experiments which fixed for all the graph instances were the probability p_N, fixed at a 0.9; λ, which was fixed to 0.1; and T_s, which was fixed to $2 \cdot |V|$.

Numerical results. There are three types of best known values (BKV) that can be found in Tables 2 and 3. The first ones are optimal results (in boldface); the second ones are values computed by heuristics and, hence, we do not know whether they are optimal or not. Finally we have instances for which only upper

Table 3. Results for the QAPlib benchmark

Graph	|V|	|E|	1D Cost	2D R	C	BKV	δ	Cost	R	C	Cost	3D R	C	D	Cost
nug12	12	45	1000	3	4	**578**	0%	578			-	2	2	3	524
nug14	14	68	1866	3	5	**1014**	0%	1014			-	2	2	4	920
nug15	15	75	2186	3	5	**1150**	0%	1150			-	2	2	4	1030
nug16a	16	93	3050	4	5	**1610**	0%	1550	4	4	1550	2	2	4	1398
nug16b	16	84	2400	4	4	**1240**	0%	1240	4	4	1240	2	2	4	1130
nug17	17	101	3388	4	5	**1732**	0%	1672	3	6	1672	2	3	3	1466
nug18	18	113	3986	4	5	**1930**	0%	1900	3	6	1900	2	3	3	1646
nug20	20	141	5642	4	5	**2570**	0%	2570			-	2	2	5	2352
nug21	21	137	5084	3	7	**2438**	0%	2438	4	6	2270	2	3	4	1988
nug22	22	153	6184	2	11	**3596**	0%	3596	4	6	2742	2	3	4	2344
nug24	24	185	8270	4	6	**3488**	0%	3488			-	2	3	4	2938
nug25	25	200	9236	5	5	**3744**	0%	3744			-	3	3	3	3100
nug27	27	233	11768	3	9	**5234**	0%	5234	5	6	4612	3	3	3	3802
nug28	28	251	13090	4	7	**5166**	0%	5166	5	6	4988	2	3	5	4302
nug30	30	293	16502	5	6	**6124**	0%	6124			-	2	3	5	5240
scr12	12	28	42776	3	4	**31410**	0%	31410			-	2	2	3	30490
scr15	15	42	80862	4	4	**51140**	0%	51140			-	2	2	4	49968
scr20	20	62	183270	5	4	**110030**	0%	110030			-	2	2	5	101686
sko100a	100	3431	757188	10	10	152002	0,016%	152026			-	4	5	5	103176
sko100b	100	3414	771792	10	10	153890	0,005%	153898			-	4	5	5	104186
sko100c	100	3372	736510	10	10	147862	0%	147862			-	4	5	5	100438
sko100d	100	3367	747542	10	10	149576	0,011%	149592			-	4	5	5	101452
sko100e	100	3366	745104	10	10	149150	0,008%	149162			-	4	5	5	101330
sko100f	100	3377	746562	10	10	149036	0,005%	149044			-	4	5	5	100922
sko42	42	603	51050	6	7	15812	0%	15812			-	2	3	7	13758
sko49	49	811	81964	7	7	23386	0%	23386			-	2	5	5	18856
sko56	56	1061	128106	7	8	34458	0%	34458			-	2	4	7	28396
sko64	64	1386	193878	8	8	48498	0%	48498			-	4	4	4	34962
sko72	72	1781	278408	8	9	66256	0%	66256			-	3	4	6	48800
sko81	81	2274	410562	9	9	90998	0%	90998			-	3	3	9	73022
sko90	90	2771	547124	9	10	115534	0%	115534			-	3	5	6	82248
ste36a	34	172	20574	2	17	**9526**	0%	9526	5	7	9258	3	3	4	8226
tho150	150	4732	48711062	10	15	8133398	0,114%	8142732	12	13	7926106	5	5	6	5088332
tho30	30	217	348124	3	10	**149936**	0%	149936	5	6	128772	2	3	5	109408
tho40	40	312	729452	4	10	240516	0%	240516	6	7	232752	2	4	5	192988
wil100	100	4459	1372700	10	10	273038	0%	273038			-	4	5	5	184756
wil50	50	1099	163508	5	10	48816	0%	48816	7	8	45672	2	5	5	37090

and lower bounds are found in the literature and are represented with a range of values. The BKVs from Table 2 come from works [28] and [30] and the upper/lower bounds from [5] and [6]. On the other hand, the BKVs from Table 3 come from [10,7,21,33].

These results show that JAM is capable of matching most of the BKVs for the evaluated instances. Moreover, JAM even improved some of the results found in [6] for some minLA instances. The remarkable aspect of matching and improving some of these results is that, while they were achieved by heuristics devoted and optimized for a particular problem, JAM is able to perform with very competitive results with benchmark instances from multiple problems and in multiple dimensions. This fact also allows us to propose different layouts, enabling extra locations, that let us find layouts for which the evaluated graphs would reduce their costs. This means that, for an unconstrained real problem, we would be able to propose a layout with more locations than facilities and aim to find the best possible arrangement.

4 Conclusions

In this paper we have presented the JAM algorithm for the Multidimensional Arrangement Problem. We have tested its practicality with benchmarks from the minLA and QAP literature. The results obtained with JAM often match the best known results and even improve some of them. Our experiments provide results for 1, 2 and 3 dimensions for 81 different graphs, broadening the available instances for both minLA and QAP as well as creating a valid set of benchmark instances for MAP.

As future work we intend to find application for JAM in real scenarios. For instance, we plan to apply JAM to find efficient deployments of data center topologies in a data center physical layout.

References

1. Aarts, E.H.L., van Laarhoven, P.J.M.: Statistical cooling: A general approach to combinatorial optimization problems. Philips Journal of Research 40(4), 193 (1985)
2. Benlic, U., Hao, J.-K.: Breakout local search for the quadratic assignment problem. Applied Mathematics and Computation 219(9), 4800–4815 (2013)
3. Burkard, R.E., Karisch, S.E., Rendl, F.: Qaplib–a quadratic assignment problem library. Journal of Global Optimization 10(4), 391–403 (1997)
4. Burkard, R.E., Rendl, F.: A thermodynamically motivated simulation procedure for combinatorial optimization problems. European Journal of Operational Research 17(2), 169–174 (1984)
5. Caprara, A., Letchford, A.N., Salazar-González, J.-J.: Decorous lower bounds for minimum linear arrangement. INFORMS Journal on Computing 23(1), 26–40 (2011)
6. Caprara, A., Oswald, M., Reinelt, G., Schwarz, R., Traversi, E.: Optimal linear arrangements using betweenness variables. Mathematical Programming Computation 3(3), 261–280 (2011)
7. Connolly, D.T.: An improved annealing scheme for the qap. European Journal of Operational Research 46(1), 93–100 (1990)
8. Drezner, Z.: Compounded genetic algorithms for the quadratic assignment problem. Oper. Res. Lett. 33(5), 475–480 (2005)
9. Fischetti, M., Monaci, M., Salvagnin, D.: Three ideas for the quadratic assignment problem. Operations Research 60(4), 954–964 (2012)
10. Fleurent, C., Ferland, J.A.: Genetic hybrids for the quadratic assignment problem. In: DIMACS Series in Mathematics and Theoretical Computer Science, pp. 173–187. American Mathematical Society (1993)
11. Fleurent, C., Glover, F.: Improved constructive multistart strategies for the quadratic assignment problem using adaptive memory. INFORMS Journal on Computing 11(2), 198–204 (1999)
12. Hansen, M.D.: Approximation algorithms for geometric embeddings in the plane with applications to parallel processing problems. In: 30th Annual Symposium on FOCS 1989, pp. 604–609. IEEE (1989)
13. James, T., Rego, C., Glover, F.: Multistart tabu search and diversification strategies for the quadratic assignment problem. IEEE Trans. on Systems, Man and Cybernetics, Part A: Systems and Humans 39(3), 579–596 (2009)

14. Zhu, J., Rui, T., Fang, H., Zhang, J., Liao, M.: Simulated annealing ant colony algorithm for qap. In: ICNC 2012, pp. 789–793 (2012)
15. Kirkpatrick, S., Gelatt Jr., D., Vecchi, M.P.: Optimization by simmulated annealing. Science 220(4598), 671–680 (1983)
16. Koopmans, T.C., Beckmann, M.: Assignment problems and the location of economic activities. Econometrica: Journal of the Econometric Society, 53–76 (1957)
17. Li, Y., Pardalos, P.M., Resende, M.G.C.: A greedy randomized adaptive search procedure for the quadratic assignment problem. Quadratic Assignment and Related Problems 16, 237–261 (1994)
18. Mcallister, A.J.: A new heuristic algorithm for the linear arrangement problem. Technical Report TR-99-126a, University of New Brunswick (1999)
19. Metropolis, N., Rosenbluth, A.W., Rosenbluth, M.N., Teller, A.H., Teller, E.: Equation of state calculations by fast computing machines. The Journal of Chemical Physics 21, 1087 (1953)
20. Mills, P., Tsang, E., Ford, J.: Applying an extended guided local search to the quadratic assignment problem. Annals of Operations Research 118(1-4), 121–135 (2003)
21. Misevičius, A.: A modified simulated annealing algorithm for the quadratic assignment problem. Informatica 14(4), 497–514 (2003)
22. Misevičius, A.: An improved hybrid genetic algorithm: new results for the quadratic assignment problem. Knowl.-Based Syst. 17(2-4), 65–73 (2004)
23. Misevičius, A.: A tabu search algorithm for the quadratic assignment problem. Comp. Opt. and Appl. 30(1), 95–111 (2005)
24. Misevičius, A.: An implementation of the iterated tabu search algorithm for the quadratic assignment problem. OR Spectrum 34(3), 665–690 (2012)
25. Nissen, V., Paul, H.: A modification of threshold accepting and its application to the quadratic assignment problem. Operations-Research-Spektrum 17(2-3), 205–210 (1995)
26. Nyberg, A., Westerlund, T., Lundell, A.: Improved discrete reformulations for the quadratic assignment problem. In: Gomes, C., Sellmann, M. (eds.) CPAIOR 2013. LNCS, vol. 7874, pp. 193–203. Springer, Heidelberg (2013)
27. Petit, J.: Experiments on the minimum linear arrangement problem. Journal of Experimental Algorithmics (JEA) 8, 2–3 (2003)
28. Rodríguez-Tello, E., Hao, J.-K., Torres-Jiménez, J.: An effective two-stage simulated annealing algorithm for the minimum linear arrangement problem. Computers & Operations Research 35(10), 3331–3346 (2008)
29. Rodriguez-Tello, E., Hao, J.-K., Torres-Jiménez, J.: A refined evaluation function for the minla problem. In: Gelbukh, A., Reyes-Garcia, C.A. (eds.) MICAI 2006. LNCS (LNAI), vol. 4293, pp. 392–403. Springer, Heidelberg (2006)
30. Safro, I., Ron, D., Brandt, A.: Graph minimum linear arrangement by multilevel weighted edge contractions. Journal of Algorithms 60(1), 24–41 (2006)
31. Sahni, S., González, T.F.: P-complete approximation problems. J. ACM 23(3), 555–565 (1976)
32. Stützle, T.: Max-min ant system for quadratic assignment problems. Technical Report Forschungsbericht AIDA-97-04, TU Darmstadt (1997)
33. Taillard, É.D.: Robust taboo search for the quadratic assignment problem. Parallel Computing 17(4-5), 443–455 (1991)
34. Taillard, É.D., Gambardella, L.M.: Adaptive memories for the quadratic assignment problems. Technical report (1997)

35. Varanelli, J.M., Cohoon, J.P.: A fast method for generalized starting temperature determination in homogeneous two-stage simulated annealing systems. Computers & Operations Research 26(5), 481–503 (1999)
36. Wang, J.-C.: Solving quadratic assignment problems by a tabu based simulated annealing algorithm. In: ICIAS 2007, pp. 75–80. IEEE (2007)
37. Wang, J.-C.: A multistart simulated annealing algorithm for the quadratic assignment problem. In: IBICA 2012, pp. 19–23. IEEE (2012)
38. Wilhelm, M.R., Ward, T.L.: Solving quadratic assignment problems by simulated annealing. IIE Transactions 19(1), 107–119 (1987)
39. Zhang, C., Lin, Z., Lin, Z.: Variable neighborhood search with permutation distance for qap. In: Khosla, R., Howlett, R.J., Jain, L.C. (eds.) KES 2005. LNCS (LNAI), vol. 3684, pp. 81–88. Springer, Heidelberg (2005)

Online Performance Measures for Metaheuristic Optimization

Kay Hamacher

Dept. of Computer Science, Dept. of Physics & Dept. of Biology,
Technical University Darmstadt, Schnittspahnstr. 10, 64287 Darmstadt, Germany
http://www.kay-hamacher.de

Abstract. (Global) optimization is one of the fundamental challenges in scientific computing. Frequently, one encounters objective functions or search space topologies that do not fulfill necessary requirements for well understood and efficient procedures like, e.g., linear programming. This methodological gap is filled by metaheuristic optimization approaches. Their search dynamics in high dimensional search spaces and for complicated objective functions is not well understood at present. In particular, the choice of parameters driving the procedures is a demanding task. In this contribution we show how insight from time series analysis help to investigate – on a pure *empirical* basis – metaheuristic schemes. Rather than deriving analytical results on convergence behavior, *ex ante*, we propose *online* observation of the search and optimization progress. To this end, we use the Detrended Fluctuation Analysis – a method from time series analysis – to investigate the search dynamics of metaheuristics as stochastic processes. We apply the proposed method to two different metaheuristic, namely differential evolution and basin hopping.

1 Introduction

From parameter fitting and model selection over molecular structure prediction to optimal control researchers and engineers face global optimization (GO) problems. Whenever the objective function is nonlinear and especially not differentiable, randomized search approaches have been suggested in the literature as a promising route [23]. The first well-known success was simulated annealing [19,6]. This approach, however, suffers from the "freezing problem": the search process gets trapped in a local optimum. Among others, basin hopping (BH) [38] and stochastic tunneling [39,14] have been suggested to tackle this particular problem.

Other metaheuristics [23] to navigate the search space are inherently parallel search techniques like genetic and evolutionary algorithms, such as differential evolution (DE) [33,7]. Previous work showed that in some applications DE converges faster and with more certainty than, e.g., Adaptive Simulated Annealing as well as the annealed Nelder & Mead approach [33]. Here, the simultaneous dynamics of several solution vectors can help to escape local minima. Although, this is intuitively correct, situations arise where even a large population size cannot guarantee desired properties as good convergence speed.

M.J. Blesa, C. Blum, and S. Voß (Eds.): HM 2014, LNCS 8457, pp. 169–182, 2014.

Now, while automated parameter tuning has a long history in the research on metaheuristics [4,24] and these algorithms were successfully applied in various domains [36,9], they are still poorly understood [41] in their behavior. In contrast, other approaches – such as branch-and-bound methods – have turned out to be more easily accessible to analytic analysis [28]. While general metaheuristics might still escape an analytics treatment, their performance and search dynamics is always accessible from a pure empirical point of view.

Our contribution in this study is the *ex post* analysis of metaheuristics while we regard them as stochastic processes that can be made accessible by tools from time series analysis. We use the Detrended Fluctuation Analysis (DFA) method to this end. We were able to show that a "super-diffusive" like behavior in visited objective function values is connected with high performance; furthermore – and from the point of view of algorithm steering more important – *suboptimal* search performance is always related to random walk like behavior and thus insufficient coverage of the search space.

2   The Global Optimization Problem

2.1   Definitions

The Global Optimization (GO) problem can be stated as the task to find a best estimator x_{best} of any of the (potentially degenerated) solutions x_* to the problem[1]

$$x_* := \operatorname*{argmin}_{x \in \mathcal{D}} E(x) \tag{1}$$

for a (continuous or discrete) objective function $E(x)$. Examples for E are the energy in physics, the path-length in the well-known Traveling Salesperson Problem, or the loss function in machine learning. E is defined on a D-dimensional space $\mathcal{D}$. The function value at the global optimum will be called $E_* := E(x_*)$ and at the best estimator x_{best} it is called $E_{best} := E(x_{best})$ in the subsequent parts of this paper.

2.2   Practical Issues in Global Optimization

For algorithm development and performance evaluation one typically uses functions that are fast to compute, but nevertheless show similar characteristics as real-world applications, such as continuity, barriers and transition states, degenerated minima, or sometimes differentiability. Two well-known test cases are discussed in Sec. 4.1.

To *empirically* assess the performance of global optimization algorithms there are several approaches discussed in the literature [37,21]:

1. One might compare alternative approaches based on the computational costs (in CPU cycles, no. of iterations and so on) it takes to determine the global

---

[1] We are only concerned with *minimization*; maximization is just a trivial mapping.

optimum of the problem of Eq. 1. However, as general GO problems are $\mathcal{NP}$-hard [18,26,15,27,22,40] such a measure is most likely in itself very costly to quantify.

2. Alternatively, researchers have introduced [13] the relative error of a suggested solution E_{best} for *given computational costs* of n iterations with respect to the known global optimum of a test function $\epsilon_{\text{rel}}(n) := \frac{E_{\text{best}}(n) - E_*}{E_*}$. This approach leads to rather pragmatic insight and does not touch the subtleties of computational complexity.

3. One might also consider alternative approaches such as the distance (in some metric) in the $\mathcal{D}$-space of the suggested solution x_{best} to the known solution x_*. If several, degenerated minima exists, on would chose the "nearest" on as reference.

In the subsequent parts of this study we will employ the ϵ_{rel} measure to assess the quality of the computed solution to the problem of Eq. 1. Note, however, that ϵ_{rel} was critically discussed [42] in the context of integer programming, where the definition of other and conceptually better quality measures is possible. Other, slightly changed variants to use function values and differences to known solutions were proposed [43]. However, all this methods necessarily need to be monotonous to our ϵ_{rel} to sensible. As we will describe below, our results show *correlations* between ϵ_{rel} and our analysis procedure. Such correlations could, however, not change qualitatively under any monotonous transformation of error measures in function values.

3   Our Contribution : Time Series Analysis for Performance Assessment of Metaheuristics

The search dynamics of a metaheuristic optimization algorithm leads to a distinct time series of values of the objective function tested or evaluated. We will call this series $\{E_g\}$ for iteration or generation no. g with a total length G. We will apply the Detrended Fluctuation Analysis (DFA) [25,5] to $\{E_g\}$. Our DFA implementation in Algorithm 1 quantifies the correlations within a time series involving an overall trend of polynomial order. We did not investigate other potential trends, such as exponential or periodic ones [17], because low-degree polynomial DFA was already sufficient for our purposes (cmp. Fig. 1).

The averaged squared fluctuation of values of the time series F^2 are related to the exponent γ in the scaling law $F^2(\Gamma) \sim \Gamma^{2-\gamma}$. Quantifying the correlations by DFA is done by computing the exponent γ of the auto-correlation function of visited objective function values – regarded as a time series

$$C(t) = \frac{1}{g-t} \sum_{\tau=0}^{g-t} (E_\tau - \langle E_g \rangle) \cdot (E_{\tau+t} - \langle E_g \rangle) \sim t^{-\gamma}$$

$$\langle E_g \rangle := \frac{1}{g} \sum_{\tau=0}^{g} E_\tau$$

Algorithm 1. Detrended Fluctuation Analysis (DFA): for time window sizes between G_s and G_e a polynomial (the trend) is fitted to the cumulative series of the input series $\{E_g\}$. The mean-square deviations $F^2(\Gamma)$ scale like $\Gamma^{2-\gamma}$.

Require: $E'_g := \sum\limits_{g'=0}^{g} (E_{g'} - \langle E_{g'} \rangle)$

$\qquad\qquad\qquad\qquad\qquad\qquad$ ▷ cumulative series of E_g with $\langle \ldots \rangle$ as expectation value
$\quad$ **for** $\Gamma = G_s$ **to** G_e **do** $\qquad\qquad\qquad\qquad\qquad$ ▷ Γ : time window length
$\qquad D :=$ decompose E'_g into $\lfloor \frac{G}{\Gamma} \rfloor$ many consecutive series
$\qquad$ **for all** d in D **do**
$\qquad\qquad p :=$ Fit polynomial of order n to data in d
$\qquad\qquad F^2{}_d^{(\Gamma)} := \frac{1}{L^{(\Gamma)}} \sum\limits_{i=1}^{L^{(\Gamma)}} (d_i - p_i)^2 \qquad$ ▷ mean-square deviation of the series d of
$\qquad\qquad\qquad\qquad\qquad\qquad$ ▷ length $L^{(\Gamma)}$ from the polynomial p; i: idx of data points in d
$\qquad F^2(\Gamma) := \frac{1}{|D|} \sum_{d \in D} F^2{}_d^{(\Gamma)}$
$\quad$ Fit $F^2(\Gamma) \sim \Gamma^{2-\gamma}$ to the list of $F^2(\Gamma)$

by disregarding trends up to a given polynomial order. Now, $\gamma \approx 0$ indicates a "super-diffusive" behavior – a dynamics that explores the function value range "fast", while $\gamma \approx 1$ indicates random-walk like behavior.

Eventually, the latter case is the worst behavior any stochastic search procedure can show: at this point no "structure" or "topology" of the objective function and its transition states are leveraged and the search amounts to random guessing. In the former case ($\gamma \approx 0$), however, the search process is sampling quite efficiently and is obviously subject to some guiding force that leads to "fast" exploration of function values – as the function values $E_{g'}$ and $E_{g''}$ at different $g' \neq g''$ are correlated.

We note in passing, that previous results for directed random walks (DRW) and their efficient dynamics [10] support our notion on the dynamics of randomized optimization algorithms — due to the fact that the network of "nearest-neighbor" local minima form structured graphs [8].

It is non-obvious, but one can perform the DFA analysis in *constant memory and constant time* even under increasing overall number of iterations G: as we seek the exponent in the scaling law $F^2(\Gamma) \sim \Gamma^{2-\gamma}$ we can always store the time series $\{E_g\}$ modulo some time scale T. Using just every T-th function value from the original series $\{E_g\}$ does not change the scaling exponent γ at all.

4 Applications of DFA for Metaheuristics

4.1 Differential Evolution (DE) as an Optimization Technique

DE seeks optima by evolving a population of solution vectors. A current population $\mathcal{P}$ of P individuals is taken over into a new generation $g \rightarrow g + 1$ under mutation, recombination, and eventually selection. Since its introduction DE has been extended both in the areas of applicability as well as in design [29,7].

Algorithm 2. Differential Evolution algorithm; here $S \in \{0; 1\}$ switches between two strategies and $0 \le \lambda \le 1$ provides for a continuum on how strong to focus the search dynamics on the vicinity of the best solution found so for x_{best}. The indices r_1, r_2, r_3, r_4, r_5 are drawn from the range $[1; P]$ and are distinct from each other. n is drawn from the uniform distribution $\mathcal{U}(1, D)$, L from the truncated Poisson distribution $\mathcal{S}(L, \nu, D) \sim \nu^L$, however, not larger than D. If $L + n > D$, we obtain the indices of the components to be taken from t modulo D. Here, n is the component index at which the original vector x_i and the new "test vector" t are "recombined". L is the number of alleles, that is entries, which are transferred from t to x_i in the next generation:

$$u := (u_1, \ldots, u_D)^T = \left(x_{i,1}, x_{i,2}, \ldots, t_n, t_{n+1}, \ldots, t_{n+L}, x_{i,n+L+1}, \ldots, x_{i,D} \right)^T$$

Here $x_{i,k}$ is the k-th entry of the vector x_i.

---

for $1 \le g \le G$ do
    for all x_i in the population $\mathcal{P}$ do
        set $t := S \cdot x_{r_1} + (S-1) \cdot (x_i + \lambda [x_{\text{best}} - x_i]) + F_1 \cdot (x_{r_2} - x_{r_3}) + F_2 \cdot (x_{r_4} - x_{r_5})$
        draw L from $\mathcal{S}(L, \nu, D)$
        draw n from $\mathcal{U}(1, D)$
        copy L entries from t to u starting at index n
        copy the other $D - L$ entries from x_i to u
        if $E(u) < E(x_i)$ then
            replace x_i by u
        if $E(u) < E_{\text{best}}$ then
            $E_{\text{best}} := E(u)$
            $x_{\text{best}} := u$

After initialization, in which all individuals are set to some random starting point in the $\mathcal{D}$-space, the generations are formed according to Algorithm 2. Here, x_{best} is our estimate of the location of the global minimum and E_{best} its respective energy. Note, that the general DE scheme of Algorithm 2 enables us to model all DE variants suggested earlier [32] by appropriate choices of internal parameters P, F_1, F_2, λ, S and ν.

In the DFA analysis of DE we will record as $\{E_g\}$ the newly generated solutions, because this is the new "information" within this generation of the population. Other strategies might analyze the time series of the best individual or of an average performance. These measures are, however, not an ideal indicator of progress: suppose one analyzes the best energy; under this analysis regime we cannot obtain any information on the population as a whole. Even if the whole population is stuck in a local minimum, or it is exploring the whole energy landscape, thus randomly guessing.

A Test Problem for DE. The Shubert test function [37,30] is one of the most frequently used test functions. It shows all characteristics listed above (continuity, differentiability, ...). Its definition reads:

Algorithm 3. Basin Hopping with an acceptance criterion based on the threshold b. $U(\boldsymbol{x})$ is a neighborhood of $\boldsymbol{x}$ based on an application-dependent metric.

Require: randomly chosen start solution $\boldsymbol{x}_0$
  $E_0 := E(\boldsymbol{x}_0)$
  for $1 \le g \le G$ **do**
    $t := $ draw from $U(\boldsymbol{x}_{g-1})$
    if $E(t) - E_{g-1} < b$ **then**   ▷ accept smaller $E(t)$ or within b
      $\boldsymbol{x}_g := t$
      $E_g = E(t)$
      if $E_g < E_{\text{best}}$ **then**
          $E_{\text{best}} := E_g$ and $\boldsymbol{x}_{\text{best}} := t$
    else
      $\boldsymbol{x}_g := \boldsymbol{x}_{g-1}$ and $E_g = E_{g-1}$

$$s(x) = \sum_{k=1}^{5} k \cdot \sin\left((k+1) \cdot x + k\right)$$

Previously, $s(x)$ was used in one dimensional optimization. However, this situation is conceptually different from the problem in two and more dimensions, due to the existence of "transition states" or saddle points in higher dimensions. This insight lead to the suggestion to abandon one dimensional test functions all together [11].

A generalized Shubert function $s_D(\boldsymbol{x})$ is used for D dimensions. For $D \ge 1$, we then define $\boldsymbol{x} = (x_1, x_2, \ldots, x_D)^T \in \mathbb{R}^D$ and

$$s_D(\boldsymbol{x}) := \prod_{i=1}^{D} s(x_i)$$

defined on the D-dimensional, real-valued $\mathcal{D}$-space. In the subsequent parts of this paper we will consider the three-dimensional Shubert function $(D = 3)$ to avoid the aforementioned problem with low dimensionality $(D = 1)$.

4.2 Basin Hopping (BH) as an Optimization Technique

BH was first successfully used in chemical physics and biomolecular structure prediction [8,38]. BH is a Markov chain Monte Carlo technique, but differs in the acceptance criterion: whereas simulated annealing [19], Glauber dynamics [2], or Metropolis sampling [20] use a criterion based on or related to a Boltzmann-factor, BH applies a binary threshold criterion. In Algorithm 3 we describe BH in more detail.

A Test Instance for BH. To illustrate the general applicability of the DFA procedure we decided to apply BH to a different optimization problem, this time a discrete one: energy ground state of spin-glasses [3] for which exact solutions are available [31]. The objective function is defined as $E(s) = \sum_{<i,j>} J_{ij} s_i s_j$. The summation $<i,j>$ includes nearest neighbors[2] and $s_i \in [+1; -1]$ are Ising spins, while the interaction parameters J_{ij} are normally distributed. Here, the neighborhood $U(s)$ of a given configuration s is defined as those configurations s' with an edit distance of one, thus $|s - s'| = \sum_{i=1}^{n} |s_i - s_i'| \overset{!}{=} 1$.

5   Experimental Results

5.1   DE – General Findings

Overall, we performed some $145,000$ simulations for a sampling of parameter settings $(F_1, F_2, \lambda, S, \nu, P)$. In general, at the maximal iteration count of $G = 600,000$ we found only weak dependency on ν. Population sizes P larger than 250 always showed inferior results; this finding can be attributed to the fact that large population sizes tend to sample the solution space more homogeneously the larger the population – effectively removing correlation from the population members – and thus the rationale of evolutionary dynamics altogether.

Furthermore, the performance dependency on F_1, F_2, and λ was small; however, there are some indications that larger values for F_1, F_2 and smaller values for λ lead to better performance in the respective error measures ϵ_{rel}. It turns out that $S = 0$ showed better performance with regard to the relative error in the final estimate of the global optimum value.

5.2   DE – On-Line Performance Measure

The DFA results, however, were unambiguous: any "super-diffusive" search as indicated by small γ_{DFA} exponents showed almost always better relative errors ϵ_{rel} in the minima found so far. This is evident from the results presented in Fig. 1.

Fig. 1 also suggests potential over-fitting for third and fourth order DFA. Typically, polynomials of higher order are able to fit better through any given number of data points. However, resulting exponents are not stable due to over-fitting. The exponents for smaller degrees (first and second order DFA) were, on the other hand, not sensitive to changing random generator seeds [data not shown]. Thus, in the subsequent parts of this work we focus exclusively on first order DFA analysis.

5.3   DE – Online Adaption of Parameter Choices

The results above from Sec. 5.2 suggest a potential adaptation mechanism [12,13]: *online* estimates of the DFA exponent of first order can be used to compare the current DE-parameter vector $(F_1, F_2, \lambda, S, \nu, P)$ to any other choice and thereby

---

[2] Here in a 2D, periodic, regular lattice of size $\sqrt{n} \times \sqrt{n}$.

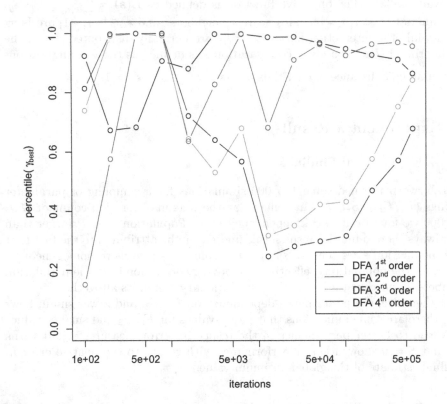

Fig. 1. For each iteration number (x-axis) the y-axis shows the percentile of sorted DFA-exponents for the sampled parameter vectors $(F_1, F_2, \lambda, S, \nu, P)$ for which the best estimator of the global optimum (the minimal ϵ_{rel}) was worse than for the estimator with the smallest γ encountered so far. Note, that a percentile close to one indicates that almost all runs showed a larger value of the scaling exponent γ. The DFA was performed up to fourth order. Over-fitting with higher polynomial order can easily be observed within the graphs.

control for suboptimal search performance. In particular non-"super-diffusive" dynamics ($\gamma \approx 1$) on the energy landscape has to be avoided.

To this end, we have revised our simulation data for the Shubert function and modeled on it an adaptive scheme: switching between periods of a) running the optimization for a particular DE-parameter vector $(F_1, F_2, \lambda, S, \nu, P)$ and b) deciding to maintain this vector or choosing a modified one; this decision is based on whether the DFA-γ indicates suboptimal search dynamics. If a switch to a new DE-parameter vector $(F_1, F_2, \lambda, S, \nu, P)$ is suggested, we choose one drawn from the range of possible values of $(F_1, F_2, \lambda, S, \nu, P)$ chosen above and present in the precomputed data set.

In comparison to previous approaches [1], that mainly focus on *a-priori* knowledge of algorithms, distributions of (local) minima and the like, our DFA-based approach is a novel strategy, that employs insights from dynamical system theory to find efficient parameters of the underlying DE algorithm.

In Figure 2a we show the results. Almost for all iteration numbers the adaptive scheme performs better than the average 25% of the best parameter choices. It never outperforms the behavior of the best parameter choice. Note, however, that the knowledge, what is the best parameter choice, is itself an *a posteriori* insight, which requires sampling over the full or a large fraction of the DE-parameter space. This would be orders of magnitude more costly in computational terms than a single run of the adaptive scheme.

To illustrate the performance increase more clearly, we show in Figure 2b the observed relative errors for the adaptive scheme and the top 25% of traditional DE runs. Within sampling errors an adaptive scheme is most likely to be better. Note, that the cases, for which the top 25% turned out to be better, are for larger ϵ_{rel} and therefore mostly for small iteration numbers and thus at the start of the optimization runs.

We conclude that using a rather simple criterion, such as a measure on the correlated dynamics in function value space E of the DE-solution vectors, can easily indicate whether a stochastic optimization protocol is exploring the search space essentially in a "random guessing" like fashion or whether the protocol exploits an underlying structure for a guided search. The proposed, very simple, adaptive scheme implements a first attempt to harness this insight. The efficiency gain is estimated to be some 100-fold as sampling dozens to hundreds of DE-parameter settings $(F_1, F_2, \lambda, S, \nu, P)$ is avoided.

5.4   BH – Results of DFA Analysis

We applied the BH algorithm to a set of 2D Ising spin glasses of size 20×20 spins and obtained the DFA exponents as described in Algorithms 3 and 1. We sampled over individual incarnations (via randomized J_{ij} choices) of spin glasses and independent runs. We repeated this sampling for systematically varied threshold parameters b. In Figs. 3a and 3b we show the empirical results. Again, we find that high DFA exponents γ go hand in hand in with inferior performance. Therefore, again the γ values can serve as *online* indicators of performance. In stark contrast to the findings for DE, however, small or vanishing γ do not necessary lead to best performance with respect to the average relative error ϵ_{rel}. We investigated this effect in more detail by increasing the number of iterations and found some (partial) explanation; our results indicate that the γ vs. ϵ_{rel} relationship evolves in the direction as in DE. Thus, we expect a more monotonous relation of ϵ_{rel} and γ the longer the run.

6   Discussion

In this paper, we have motivated the usage of time series analysis techniques to evaluate the performance of otherwise very efficient and robust stochastic optimization protocols, namely differential evolution (DE) and basin hopping (BH).

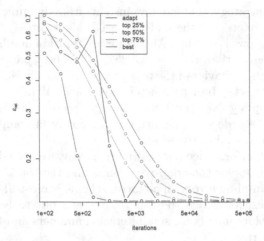

(a) The relative error ϵ_{rel} as a function of no. of iterations for various scenarios: the best parameter vector $(F_1, F_2, \lambda, S, \nu, P)$ for each iteration (blue); the averaged ϵ_{rel} for the best 25% (orange), 50% (red), and 75% (brown) of parameter vectors $(F_1, F_2, \lambda, S, \nu, P)$ at each iteration. The black line shows the results for the suggested adaptive scheme.

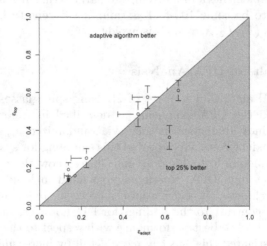

(b) Comparing the random seed averaged relative error for the top 25% of parameter vectors with the relative error for the adaptive scheme. The non-shaded area shows the regime in which the adaptive algorithm showed superior performance with regard to the $\epsilon_{\text{rel}}(n)$ criterion, while the points in the shaded area illustrate the cases in which the top 25% of parameter choices performed better. Error bars are standard deviations over the random seeds and their respective runs and the various parameter choices (for the top 25% case).

Fig. 2.

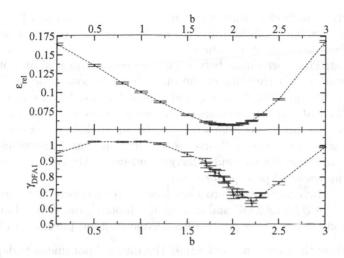

(a) Varying the threshold b in BH of Algorithm 3 affects both the DFA scaling exponent γ and thus the auto-correlation of proposed function minima, as well as the relative error ϵ_{rel} for the test case of Ising spin glasses.

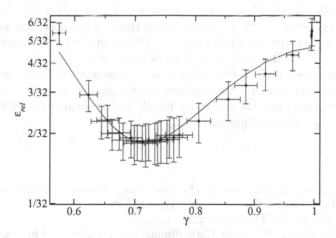

(b) Scatter-plot of data in Fig. 3a. Clearly, large values of the DFA-exponent γ lead to insufficient errors in the obtained minimal function values. Error bars are the standard deviations from the sample of test instances (x direction) and from the fitting procedure of the γ-scaling law (y direction). Black: $G = 2 \cdot 10^8$ iterations (sample of $1,000$ runs), red: $G = 3 \cdot 10^8$ iterations (sample of 400 runs).

Fig. 3.

A very instructive method of time series analysis is the detrended fluctuation analysis (DFA) paradigm, which is implemented to observe correlated searches in the space of objective function values.

We found a striking correlation between a) the performance of an optimization run in terms of relative error of the objective function value and b) the DFA exponents that quantify the dynamical regime DE and BH are working in. Typically, for the test functions employed here we found that runs with smaller DFA exponents γ performed superior when judged under their respective $\epsilon_{rel}(n)$. Such small γ represent a "super-diffusive" search dynamics and thus efficient exploitation of structure in the particular energy landscape. Always, we found that large γ are an indicator of poor performance.

Conceptually, there are two approaches how to incorporate the knowledge generated from the DFA into DE and making it adaptive, and thus a block-box algorithm without any need for *a priori* knowledge on good parameter choices:

1. we can analyze the exponents and adjust the internal parameters adaptively to increase the efficiency of the DE protocol, or/and
2. we can take the stochastic behavior as an indicator, when the injection of a randomly generated, new individual might push the DE process out of some trapping minimum.

The second approach constitutes an adaptive variant open for further investigation and beyond the scope of this study. The first variant was investigated and shown to eventually make this DE-enhancement parameter-free. We hope that this study will encourage further research along these lines and encourage other researchers to improve our knowledge not only about convergence properties and the like, but also about dynamical behavior of algorithms [41].

Note, that previous investigation of heuristics under their "time behavior" [16,34,35] measure the *overall* time of convergence and their respective distributions. Therefore, these approaches would hardly be applicable for online measurements, e.g., in *online* parameter tuning.

References

1. Bäck, T., Schwefel, H.: An overview of evolutionary algorithms for parameter optimization. Evolutionary Computation 1(1), 1–23 (1993), http://dx.doi.org/10.1162/evco.1993.1.1.1
2. Binder, K., Heermann, D.: Monte Carlo Simulation in Statistical Physics, 3rd edn. Springer, Berlin (1997)
3. Binder, K., Young, A.: Spin glasses: Experimental facts, theoretical concepts, and open questions. Rev. Mod. Phys. 58(4), 801–976 (1986)
4. Birattari, M.: Tuning Metaheuristics. SCI, vol. 197. Springer, Heidelberg (2009)
5. Bunde, A., Kantelhardt, J.: Langzeitkorrelationen in der natur: von klima, erbgut und herzrhythmus. Phys. Bl. 57(5), 49–54 (2001)
6. Chou, C., Hand, R., Li, S., Lee, T.: Guided simulated annealing method for optimization problems. Phys. Rev. E 67, 66704 (2003)
7. Das, S., Suganthan, P.: Differential evolution: A survey of the state-of-the-art. IEEE Transactions on Evolutionary Computation 15(1), 4–31 (2011)

8. Doye, J., Wales, D.: Saddle points and dynamics of Lennard-Jones clusters, solids, and supercooled liquids. J. Chem. Phys. 116(9), 3777–3788 (2002)
9. Friedrich, T., Kroeger, T., Neumann, F.: Weighted preferences in evolutionary multi-objective optimization. In: Wang, D., Reynolds, M. (eds.) AI 2011. LNCS, vol. 7106, pp. 291–300. Springer, Heidelberg (2011)
10. Friedrich, T., Sauerwald, T.: The cover time of deterministic random walks. In: Thai, M.T., Sahni, S. (eds.) COCOON 2010. LNCS, vol. 6196, pp. 130–139. Springer, Heidelberg (2010)
11. Hamacher, K.: On stochastic global optimization of one-dimensional functions. Physica A 354, 547–557 (2005)
12. Hamacher, K.: Adaptation in stochastic tunneling global optimization of complex potential energy landscapes. Europhys. Lett. 74(6), 944–950 (2006)
13. Hamacher, K.: Adaptive extremal optimization by detrended fluctuation analysis. J. Comp. Phys. 227(2), 1500–1509 (2007)
14. Hamacher, K., Wenzel, W.: The scaling behaviour of stochastic minimization algorithms in a perfect funnel landscape. Phys. Rev. E 59(1), 938–941 (1999)
15. Hansmann, U., Wille, L.T.: Global Optimization by Energy Landscape Paving. Phys. Rev. Lett. 88(23), 68105 (2002)
16. Hoos, H., Stützle, T.: On the empirical evaluation of Las Vegas algorithms (1998)
17. Hu, K., Ivanov, P.C., Chen, Z., Carpena, P., Eugene Stanley, H.: Effect of trends on detrended fluctuation analysis. Phys. Rev. E 64(1), 011114 (2001)
18. Jack, W., Rogers, J., Donnelly, R.A.: Potential transformation methods for large-scale global optimization. SIAM Journal on Optimization 5(4), 871–891 (1995), http://link.aip.org/link/?SJE/5/871/1
19. Kirkpatrick, S., Gelatt, C., Vecchi, M.: Optimization by simulated annealing. Science 220, 671–680 (1983)
20. Metropolis, N., Rosenbluth, A.W., Rosenbluth, M.N., Teller, A.H., Teller, E.: Equation of state calculations by fast computing machines. J. Chem. Phys. 21(6), 1087–1092 (1953)
21. Panos, M., Pardalos, D.S., Xue, G. (eds.): Global Minimization of Nonconvex Energy Functions: Molecular Conformation and Protein Folding. dIMACS workshop, March 20-21. DIMACS – Series in Discrete Mathematics and Theoretical Computer Science, vol. 23 (1995)
22. Pardalos, P.M., Shalloway, D., Xue, G.: Optimization methods for computing global minima of nonvoncex potential energy functions. J. Glob. Opt. 4, 117–133 (1994)
23. Pardalos, P., Romeijn, E., Tuy, H.: Recent developments and trends in global optimization. J. Comp. Appl. Math. 124(1-2), 209–228 (2000)
24. Pellegrini, P., Stützle, T., Birattari, M.: Off-line vs. on-line tuning: A study on MAX-MIN ant system for the TSP, pp. 239–250 (2010)
25. Peng, C.K., Buldyrev, S., Havlin, S., Simons, M., Stanley, H., Goldberger, A.: Mosaic organization of dna nucleotides. Phys. Rev. E 49, 1685 (1994)
26. Ratschek, H., Rokne, J.G.: Efficiency of a global optimization algorithm. SIAM Journal on Numerical Analysis 24(5), 1191–1201 (1987), http://link.aip.org/link/?SNA/24/1191/1
27. Schelstraete, S., Schepens, W., Verschelde, H.: Energy minimization by smoothing techniques: a survey. In: Balbuena, P., Seminario, J. (eds.) Molecular Dynamics: From Classical to Quantum Methods, Amsterdam, pp. 129–185 (1999)
28. Schöbel, A., Scholz, D.: The theoretical and empirical rate of convergence for geometric branch-and-bound methods. J. Global Optimization 48(3), 473–495 (2010)

29. Shi, Y.-j., Teng, H.-f., Li, Z.-q.: Cooperative co-evolutionary differential evolution for function optimization. In: Wang, L., Chen, K., S. Ong, Y. (eds.) ICNC 2005. LNCS, vol. 3611, pp. 1080–1088. Springer, Heidelberg (2005)

30. Shubert, B.O.: A sequential method seeking the global maximum of a function. SIAM J. Numer. Anal. 9(3), 379–388 (1972)

31. Simone, C., Diehl, M., Jünger, M., Mutzel, P., Reinelt, G.: Exact ground states of ising spin glasses: New experimental results with a branch-and-cut algorithm. J. Stat. Phys. 80, 487 (1995)

32. Storn, R.: On the usage of differential evolution for function optimization. In: 1996 Biennial Conference of the North American Fuzzy Information Processing Society (1996)

33. Storn, R., Price, K.: Differential evolution – a simple and efficient heuristic for global optimization over continuous spaces. J. Glob. Opt. 11(4), 341–359 (1997)

34. Stützle, T.: Iterated local search for the quadratic assignment problem. European Journal of Operational Research 174(3), 1519–1539 (2006)

35. Sttzle, T., Hoos, H.H.: Analyzing the run-time behaviour of iterated local search for the TSP. In: III Metaheuristics International Conference. Kluwer Academic Publishers (1999)

36. Sutton, A.M., Neumann, F.: A parameterized runtime analysis of evolutionary algorithms for the euclidean traveling salesperson problem. In: Hoffmann, J., Selman, B. (eds.) AAAI, AAAI Press (2012)

37. Törn, A., Žilinskas, A.: Global Optimization. LNCS, vol. 350. Springer, Heidelberg (1989)

38. Wales, D.J., Scheraga, H.A.: Global Optimization of Clusters, Crystals, and Biomolecules. Science 285(5432), 1368–1372 (1999), http://www.sciencemag.org/cgi/content/abstract/285/5432/1368

39. Wenzel, W., Hamacher, K.: A Stochastic tunneling approach for global minimization. Phys. Rev. Lett. 82(15), 3003–3007 (1999)

40. Wolpert, D.H., Macready, W.G.: No free lunch theorems for optimization. IEEE Transactions on Evolutionary Computation 1(1), 67–82 (1997)

41. Yang, X.S.: Metaheuristic optimization: Algorithm analysis and open problems. In: Pardalos, P.M., Rebennack, S. (eds.) SEA 2011. LNCS, vol. 6630, pp. 21–32. Springer, Heidelberg (2011)

42. Zemel, E.: Measuring the quality of approximate solutions to zero-one programming problems. Mathematics of Operations Research 6(3), 319–332 (1981)

43. Zlochin, M., Dorigo, M.: Model-based search for combinatorial optimization: A comparative study. In: Guervós, J.J.M., Adamidis, P.A., Beyer, H.-G., Fernández-Villacañas, J.-L., Schwefel, H.-P. (eds.) PPSN 2002. LNCS, vol. 2439, pp. 651–661. Springer, Heidelberg (2002)

Speeding Up Logic-Based Benders' Decomposition by a Metaheuristic for a Bi-Level Capacitated Vehicle Routing Problem*

Günther R. Raidl[1], Thomas Baumhauer[1,2], and Bin Hu[1]

[1] Institute of Computer Graphics and Algorithms,
Vienna University of Technology, Vienna, Austria
{raidl,hu}@ads.tuwien.ac.at
[2] Salzburg Research, Salzburg, Austria
thomas.baumhauer@gmail.com

Abstract. Benders' Decomposition (BD) is a prominent technique for tackling large mixed integer programming problems having a certain structure by iteratively solving a series of smaller master and subproblem instances. We apply a generalization of this technique called Logic-Based BD, which does not restrict the subproblems to have continuous variables only, to a bi-level vehicle routing problem originating in the timely distribution of printed newspapers to subscribers. When solving all master and subproblem instances exactly by CPLEX, it turns out that the scalability of the approach is quite limited. The situation can be dramatically improved when using a meaningful metaheuristic – in our case a variable neighborhood search – for approximately solving either only the subproblems or both, the master as well as the subproblem instances. More generally, it is shown that Logic-Based BD can be a highly promising framework also for hybrid metaheuristics.

1 Introduction

Benders' Decomposition (BD) [1] is a classical and frequently applied approach for solving large *Mixed Integer Linear Programming* (MIP) problems having a special block-diagonal structure with "complicating variables". It essentially reformulates a given problem by expressing it as a master problem on only a subset of all original variables – the complicating ones – and considering the contributions of all further variables by additional inequalities, so-called *Benders' cuts*. The optimization starts by solving a restricted form of the master problem without any or with only few of these inequalities. A new Benders' cut is then identified by solving a subproblem and its dual on the remaining variables with the master problem variables fixed to the current master solution. Obtained Benders' cuts are added to the master problem and the process is iterated until no further Benders' cuts can be derived. When the master problem and all subproblems are solved to optimality, the finally obtained solution also is optimal for the original problem.

---

* This work is supported by the Austrian Ministry for Transport, Innovation and Technology, the Ferderal State of Salzburg, Austria, and the Austrian Science Fund (FWF) under grant P24660.

M.J. Blesa, C. Blum, and S. Voß (Eds.): HM 2014, LNCS 8457, pp. 183–197, 2014.

A major restriction of this original form of BD is the fact that the subproblem must be a *Linear Programming* (LP) problem with only continuous variables as its dual solution is required to derive the Benders' cuts. Some authors, however, have also generalized BD to other types of subproblems, such as certain kinds of continuous non-linear ones [2]. In particular, Hooker and Ottosson [3] proposed *logic-based BD*, which is applicable to a wide category of subproblems including discrete ones. This is achieved by generalizing the LP dual to an *inference dual*. Constraint programming techniques turned out to be especially useful in conjunction with logic-based BD, and this combination could be successfully applied to several problems, in particular in the planning and scheduling domain [4].

In other works it has been shown that metaheuristics can be very useful in conjunction with classical BD: While the LP subproblems are usually solved efficiently by an LP-solver, the master problem typically remains a MIP, although smaller than the original problem, and in general needs to be resolved with newly added Benders' cuts many times. It has therefore been suggested to solve the master problem only approximately but faster by means of metaheuristics, and possibly only in the end apply an exact method in order to obtain a guaranteed optimum. Poojari and Beasley [5] describe such an approach for solving general MIPs in which a genetic algorithm together with a feasibility pump heuristic are applied to the master problem. The authors argue that a population based metaheuristic like a genetic algorithm is particularly useful as it provides multiple solutions in each iteration giving rise to more Benders' cuts. Similarly in spirit, Lai et al. [6, 7] propose a genetic algorithm/BD hybrid for solving the capacitated plant location problem; results indicate a tremendous saving of computation time in comparison to classical BD. Lai et al. [8] further discuss such an approach for a *Capacitated Vehicle Routing Problem* (VRP). Rei et al. [9] suggest to use local branching for solving a MIP master problem in order to sooner find improved upper as well as lower bounds.

It has also been recognized that BD subproblems need not necessarily always to be solved to optimality in order to obtain useful Benders' cuts, even when completeness of the whole approach shall be retained [10]. Especially when considering difficult subproblems in logic-based BD, this aspect becomes increasingly interesting. However, we are not aware of any work so far where metaheuristics have been applied to discrete BD subproblems for deriving Benders' cuts. The major reason obviously lies in the difficulty that it is not sufficient to find a heuristic solution to the subproblem but dual solution information is also required for identifying Benders' cuts that are guaranteed to be valid for any master problem solution. In fact, suboptimal solutions to the subproblem may easily yield inequalities that cut away too large portions of the search space, possibly also a global optimum.

This work considers a bi-level vehicle routing problem motivated by the time-critical distribution of newspapers from a printing center via satellite depots to subscribers and demonstrates how a metaheuristic may effectively be applied to the master as well as the subproblem instances of a suitable logic-based BD. Experimental results indicate that high-quality solutions can be obtained much faster than when using CPLEX for solving the master and subproblem instances exactly, and the scalability of the BD to large instances is substantially improved.

The next Sections 2 to 4 introduce the considered bi-level vehicle routing problem, refer to related work, and present a basic MIP formulation, respectively. Section 5 describes the applied logic-based BD. All the metaheuristic enhancements are presented in Section 6. Experimental results of the basic MIP, classical logic-based BD where all subproblems are solved to optimality, and metaheuristic hybrid variants are discussed in Section 7. Finally, Section 8 concludes this article with remarks on future work.

2 The Bi-Level Capacitated Vehicle Routing Problem with Time Limits

We consider a two-level vehicle routing problem in which goods shall be transported from a main depot to satellite depots and from there further to customers. Homogeneous vehicle fleets exist at the main depot and each satellite depot. A global time limit is imposed on all deliveries, i.e., each customer has to receive its goods within this time. In contrast to the two-echelon vehicle routing problem known in the literature [11–13], the assignment of customers to the satellite depots is pre-specified in our case.

This problem is motivated by the real-world scenario at Mediaprint, a major Austrian newspaper print shop who has to distribute printed newspapers from each printing center to subscribers within a guaranteed time. A natural assignment of subscribers to satellite depots arises here from the fact that region-specific supplements such as advertisements are added to the newspapers, and each region-specific version is only distributed via a dedicated satellite depot. The real distribution scenario even comprises three levels, but it turns out that only the first two levels, up to certain delivery points we call customers here, can be meaningfully optimized as the lowest level corresponds to routes of delivery agents who do not need a more serious planning or do this on their own.

We define the *Bi-Level Capacitated Vehicle Routing Problem with Time Limits* (2L-VRP-TL) as follows. Given are

- a complete, directed graph $G_0 = (V_0, A_0)$ with node set $V_0 = \{0\} \cup V_0'$ and arc set $A_0 = V_0 \times V_0$, where the special node 0 represents a main depot and V_0' the set of further satellite depots;
- for each satellite depot $s \in V_0'$ a complete, directed graph $G_s = (V_s, A_s)$ with node set $V_s = \{s\} \cup V_s'$, where V_s' represents a set of customers that receive their deliveries via satellite depot s;
- a demand $q_v \geq 0$ for each customer $v \in V_s'$ and a resulting total demand $q_s = \sum_{v \in V_s'} d_v$ for each satellite depot $s \in V_0'$;
- travel cost $c_{u,v} \geq 0$ and a travel time $t_{u,v} \geq 0$ for each arc $(u, v) \in \bigcup_{s \in V_0} A_s$ representing the fastest way to go from u to v;
- vehicle capacities $Q_s \geq 0$ for each vehicle starting at depot $s \in V_0$; thus, we assume a homogeneous vehicle fleet for each depot and the number of vehicles is not limited; in our practical application, larger vehicles are used for the first level and smaller ones for the second level;
- and a global time limit T (due time) within which all deliveries at customers have to be performed.

A solution R consists of a set of routes R_s in each subgraph G_s, $\forall s \in V_0$, with a route $r \in R_s$ being an ordered sequence of nodes $r = (r_i)_{i=1,\dots,|r|}$ with $r_i \in V'_s$. Each vehicle starts its route at the depot s, visits the nodes as specified by r and finally has to return to the depot again. For convenience, we also define $r_0 = r_{|r|+1} = s$. Each node except the main depot 0 has to be visited exactly once, all satellite depots within the first-level routes R_0 and all customer nodes within the second-level routes $\bigcup_{s \in V'_0} R_s$. Thus, each set of routes R_s also defines a partitioning of V'_s.

The cost $c(r)$ of a route $r \in R_s$, $\forall s \in V_0$, is

$$c(r) = \sum_{i=1}^{|r|+1} c_{r_{i-1}, r_i}, \tag{1}$$

the route's total demand is

$$q(r) = \sum_{i=1}^{|r|} q_{r_i}, \tag{2}$$

and the times needed to reach each node r_i from the route's depot s are

$$t(r_i) = \sum_{j=1}^{i} t_{r_{j-1}, r_j} \quad \forall i = 1, \dots, |r|. \tag{3}$$

A solution is feasible if the routes satisfy the capacity constraints

$$q(r) \le Q_s \quad \forall r \in R_s, s \in V_0, \tag{4}$$

and all deliveries are performed within the due time T. Since the second-level tours may only start after the goods have been delivered to the respective satellite depots by the first-level tours, the latter holds when

$$t(s) + t(v) \le T \quad \forall v \in V'_s, \ s \in V'_0. \tag{5}$$

The objective is to minimize the total cost of a solution, which is the sum over all its routes' costs

$$c(R) = \sum_{s \in V_0} \sum_{r \in R_s} c(r). \tag{6}$$

3   Related Work

As already mentioned, 2L-VRP-TL is related to the *Two-Echelon Vehicle Routing Problem* (2E-VRP) [11], in which also a two-level distribution via satellite depots is considered. Major differences are, however, that in 2E-VRP no time constraints are considered and the assignments of customers to satellites are not fixed but shall also be optimized. This additional degree of flexibility makes 2E-VRP even harder to solve in practice. Perboli et al. [11] propose a flow-based MIP model, strengthening inequalities, and two matheuristics. Experimental results are shown for instances with up to 50 customers and four satellites.

Crainic et al. [14] describe for the same problem multi-start heuristics based on separating the depot-to-satellite transfer and the satellite-to-customer delivery by iteratively solving the two resulting routing problems. In its spirit, this concept comes close to our logic-based Benders' decomposition, although it is not an exact approach. Hemmelmayr et al. [12] further describe an adaptive large neighborhood search heuristic involving several neighborhood structures exploiting specificities of the 2E-VRP.

Already in 1989, Jacobsen and Madsen [15] addressed the *Two-Echelon Location-Routing Problem* in the context of newspaper delivery, which further generalizes 2E-VRP by the additional aspect of deciding at which locations to open facilities (corresponding to depots). The authors suggest and compare three rather simple construction heuristics. Later more sophisticated approaches include a tabu search [16], diverse MIP models [17], and a variable neighborhood search [13].

Concerning BD and more classical (single-level) VRPs, Fisher and Jaikumar [18] describe an approach where the master problem is a general assignment problem and the subproblem is a traveling salesman problem with time-windows for each vehicle. Lai et al. [8] propose the already mentioned hybrid of BD and a genetic algorithm. Here the VRP is expressed by a multi-commodity flow formulation, the subproblems are network flow problems that can be solved efficiently, and the remaining master problem is approximately solved by the genetic algorithm.

For a more general introduction that presents BD and Lagrangian relaxation from a metaheuristic design perspective see [19].

4   MIP Model for 2L-VRP-TL

The above introduced 2L-VRP-TL can be modeled by the following MIP using variables

- $x_{u,v} \in \{0,1\}$, $\forall (u,v) \in A_s$, $s \in V_0$ indicating the arcs used for realizing the routes and
- $t_v \geq 0$, $\forall v \in V_s$, $s \in V_0$ corresponding to the above defined $t(v)$, i.e., the time needed make the delivery at v from starting at the respective depot s.

(2L-VRP-TL)

$$\text{minimize} \quad \sum_{s \in V_0} \sum_{(u,v) \in A_s} c_{u,v}\, x_{u,v} \tag{7}$$

$$\text{s.t.} \quad (x(A_s), t(V_s')) \in \text{VRP}(G_s) \qquad \forall s \in V_0 \tag{8}$$

$$t_s + t_v \leq T \qquad \forall v \in V_s,\, s \in V_0' \tag{9}$$

$$0 \leq t_v \leq T \qquad \forall v \in V_s',\, s \in V_0 \tag{10}$$

$$x_{u,v} \in \{0,1\} \qquad \forall (u,v) \in A_s,\, s \in V_0 \tag{11}$$

In (8) $\text{VRP}(G_s)$ represents a valid formulation for the classical capacitated vehicle routing problem including the calculation of the corresponding traveling times $t(v)$ on graph G_s expressed on the variables $x(A_s)$ and $t_v(V_s)$. Equations (9) limit the total times for the deliveries at all customers to T.

VRP(G_s), for $s \in V_0$, can be expressed in different ways, for simplicity we use here the following compact Miller-Tucker-Zemlin-based formulation, see e.g. [20], although significantly fore effective (but much more complex) approaches exist. Additionally used variables are

- $g_v \geq 0, v \in V_s$ corresponding to the total demand of the nodes in the tour starting from s up to (and including) v.

(VRP(G_s))

$$\sum_{v \in V_s} x_{u,v} = 1 \qquad\qquad \forall u \in V_s' \qquad (12)$$

$$\sum_{u \in V_s} x_{u,v} = 1 \qquad\qquad \forall v \in V_s' \qquad (13)$$

$$\sum_{v \in V_s'} x_{s,v} = \sum_{u \in V_s'} x_{u,s} \qquad\qquad (14)$$

$$g_v - g_u + Q_s(1 - x_{u,v}) \geq q_v \qquad \forall (u,v) \in A_s,\, u \neq s,\, v \neq s \qquad (15)$$

$$g_v + q_v(1 - x_{s,v}) \geq q_v \qquad\qquad \forall (s,v) \in A_s \qquad (16)$$

$$t_v - t_u + (T + t_{u,v})(1 - x_{u,v}) \geq t_{u,v} \qquad \forall (u,v) \in A_s,\, u \neq s,\, v \neq s \qquad (17)$$

$$t_v + t_{u,v}(1 - x_{s,v}) \geq t_{u,v} \qquad\qquad \forall (s,v) \in A_s \qquad (18)$$

$$0 \leq g_u \leq Q_s \qquad\qquad \forall u \in V_s' \qquad (19)$$

Inequalities (12) and (13) state that any node other than s must have exactly one ingoing and one outgoing arc. Equality (14) ensures that every tour must finish at s or more precisely that s has the same number of ingoing and outgoing arcs. Inequalities (15) and (16) are the Miller-Tucker-Zemlin constraints that calculate the amounts of goods delivered up to node v. The domains of variables g_v (19) ensure that the capacity Q_s of a vehicles is not exceeded. Likewise inequalities (17) and (18) are used to calculate the traveling times up to each node v as defined above.

We can further strengthen VRP(G_s) by the following inequalities from [20]:

$$Q_s \sum_{u \in V_s'} x_{u,s} \geq \sum_{v \in V_s'} q_v \qquad\qquad (20)$$

$$\sum_{u,v \in U,\, u \neq v} x_{u,v} \leq |U| - \left\lceil \frac{\sum_{u \in U} q_u}{Q_s} \right\rceil \qquad \forall U \subseteq V_s' \qquad (21)$$

In our implementation we initially provide inequalities (21) for subsets U of cardinality two and three, but do not separate the more general ones as cuts.

5   Logic-Based Benders' Decomposition for 2L-VRP-TL

Hooker [3] generalized classical BD to logic-based BD by replacing the LP dual with a so-called *inference dual*. Benders' cuts need not to be linear inequalities anymore but are more general functions. Benders' subproblems may then involve discrete variables and nonlinear functions.

We apply this approach here and decompose the above MIP model for 2L-VRP-TL into a master problem corresponding to the first-level VRP augmented with Benders' cuts and a Benders' subproblem that decouples into a set of $|V_0'|$ independent second-level VRPs. More specifically, our *master problem* is

(MP)

$$\text{minimize} \quad \sum_{(u,v)\in A_0} c_{u,v}\, x_{u,v} + \sum_{s\in V_0'} c_s \tag{22}$$

$$\text{s.t.} \quad (x(A_0), t(V_0')) \in \text{VRP}(G_0) \tag{23}$$

$$c_s \geq \beta_{t_s^k}(t_s) \qquad\qquad k \in K_s,\ s \in V_0' \tag{24}$$

$$0 \leq t_s \leq T \qquad\qquad \forall s \in V_0' \tag{25}$$

$$0 \leq c_s \qquad\qquad \forall s \in V_0' \tag{26}$$

$$x_{u,v} \in \{0,1\} \qquad\qquad \forall (u,v) \in A_0 \tag{27}$$

It only considers the first-level decision variables $x_{u,v}$ and t_s associated with G_0 and new variables c_s representing (upper bounds for) the total cost of the second-level tours in G_s for each satellite depot s. Inequalities (24) are the Benders' cuts relating c_s with t_s in order to ultimately ensure optimality.

The associated *Benders' subproblem* to be solved for deriving Benders' cuts assumes the above master problem variables t_s to be fixed to some current values t_s^k and becomes for each $s \in V_0'$

$(\text{SP}_s(t_s^k))$

$$\text{minimize} \quad \sum_{(u,v)\in A_s} c_{u,v}\, x_{u,v} \tag{28}$$

$$\text{s.t.} \quad (x(A_s), t(V_s')) \in \text{VRP}(G_s) \tag{29}$$

$$0 \leq t_v \leq T - t_s^k \qquad\qquad \forall v \in V_s' \tag{30}$$

$$x_{u,v} \in \{0,1\} \qquad\qquad \forall (u,v) \in A_s \tag{31}$$

Thus, a minimum cost VRP-solution on G_s with delivery times at most $T - t_s^k$ shall be found for each $s \in V_0'$.

In general, Benders' algorithm starts by solving MP with none or only a small set of initial Benders' cuts. This yields values for the MP variables, i.e., t_s^k, for which the subproblem and its dual are solved in order to derive one or more cuts. These are added to the MP and the whole process is iterated until no further violated cuts exist. It has been shown that when the master problem as well as the duals and the associated inference duals are always solved to optimality, an optimal solution for the original problem will be obtained [3].

The inference dual of subproblem $(\text{SP}_s(t_s^k))$ is

$(\mathrm{DSP}_s(t_s^k))$

maximize β_s (32)

s.t. $(x(A_s), t(V_s')) \in \mathrm{VRP}(G_s) \wedge (t_v \leq T - t_s^k \; \forall v \in V_s')$

$$\xrightarrow{\{0,1\}^{|A_s|}, [0,T]^{|V_s'|}} \sum_{(u,v) \in A_s} c_{u,v} \, x_{u,v} \geq \beta_s \qquad (33)$$

i.e., to find the best possible lower bound β_s on the cost $\sum_{(u,v) \in A_s} c_{u,v} \, x_{u,v}$ that can be inferred from the constraints, assuming the fixed t_s^k.

The heart of Logic-based Benders' decomposition is now to derive from this result a more general function $\beta_{t_s^k}(t_s)$ that gives a valid lower bound on the optimal value of the cost $\sum_{(u,v) \in A_s} c_{u,v} \, x_{u,v}$ for any given value of t_s, ideally with $\beta_{t_s^k}(t_s^k)$ corresponding to the optimal solution value of $\mathrm{SP}_s(t_s^k)$. This function $\beta_{t_s^k}(t_s)$ then directly yields a corresponding Benders' cut (24).

Fortunately, in our case the situation is relatively simple. We can observe that increasing or decreasing t_s^k results in stronger or weaker constraints for SP_s, respectively, and consequently the subproblem's cost will weakly monotonically increase with t_s^k. Consider a current $\mathrm{SP}_s(t_s^k)$ and assume it is bounded and non-empty and thus has an optimal solution. Let c_s^k be this solution's cost. From the previous observations we can define a Benders' cut

$$c_s \geq \beta_{t_s^k}(t_s) = \begin{cases} c_s^k & \text{if } t_s \geq t_s^k \\ 0 & \text{else.} \end{cases} \qquad (34)$$

Intuitively this means that the costs are at least c_s^k or we need more than $T - t_s^k$ time for the subproblem, i.e., $c_s \geq c_s^k \vee t_s < t_s^k$. As we want to solve the MP by a MIP-solver, this logic-based Benders' cut is translated into the following pair of linear inequalities

$$c_s \geq c_s^k \chi_k \qquad (35)$$

$$t_s \leq (t_s^k - \varepsilon)(1 - \chi_k) \qquad (36)$$

with $\chi_k \in \{0, 1\}$ being a new decision variable that is also added to MP and ε being a small constant to ensure $c_s \geq c_s^k$ in case of $t_s = t_s^k$.

In general, it might happen that $\mathrm{SP}_s(t_s^k)$ is infeasible. Then, $\mathrm{DSP}_s(t_s^k)$ is unbounded and a *feasibility cut* – in contrast to above *optimality cut* – needs to be derived, which is a condition that cuts away the current t_s^k from MP. In our case, however, we avoid infeasible subproblems by initially determining a minimum time required for each subproblem SP_s, $s \in V_0'$ to be solvable and limiting t_s correspondingly. As we can safely assume that the triangle inequality holds for travel times, a minimum time solution is achieved by visiting each customer by an individual vehicle directly from the depot, i.e.,

$$t_s \leq T - \max_{v \in V_s'} t_{s,v} \quad \forall s \in V_0'. \qquad (37)$$

To start with a more meaningful initial MP, general lower bounds for the subproblem costs c_s are determined by solving $\mathrm{SP}_s(0)$ and requiring $c_s \geq c_s^k \; \forall s \in V_0'$.

To avoid unnecessary recalculations, we store all solved subproblems with their op-timal solutions $(t_s^k, x^k, c_s^k, \hat{t}_s^k)$, $\forall k \in K_s$, $s \in V_0'$, with

$$\hat{t}_s^k = T - \sum_{(u,v) \in A_s, \, v \neq s} t_{u,v} \, x_{u,v} \tag{38}$$

being the latest possible time for t_s for which this subproblem solution x^k would still be feasible and optimal; note that $t_s^k \leq \hat{t}_s^k$. A new subproblem $SP_s(t_s^l)$ only needs to be processed if there is no stored solution with $t_s^k \leq t_s^l \leq \hat{t}_s^k$.

When solving $SP_s(t_s^l)$, a possibly existing record $(t_s^k, x^k, c_s^k, \hat{t}_s^k)$ with the largest t_s^k less than t_s^l yields a lower bound on the costs and a possibly existing record $(t_s^{k'}, x^{k'}, c_s^{k'}, \hat{t}_s^{k'})$ with the smallest t_s^k larger than t_s^l yields an upper bound, i.e.,

$$c_s^k \leq c_s^l \leq c_s^{k'} \tag{39}$$

can be added as strengthening inequalities, and $x^{k'}$ can be used as initial heuristic solu-tion to speed up the optimization.

Finally, when the solution x^l to $SP_s(t_s^l)$ has cost c^l that were already encountered at an earlier instance $SP_s(t_s^k)$, i.e., $\exists k \in K_s \mid c_s^l = c_s^k$, the corresponding records can be merged to

$$(\min(t_s^k, t_s^l), x^k, c_s^k, \hat{t}_s^k) \quad \text{if } \hat{t}_s^k \geq \hat{t}_s^l \tag{40}$$

$$(\min(t_s^k, t_s^l), x^l, c_s^k, \hat{t}_s^l) \quad \text{else,} \tag{41}$$

and the already existing Benders' cut $c_s \geq \beta_{t_s^k}(t_s)$ is adapted (lifted) accordingly with-out introducing a new cut.

6 Metaheuristic Improvements

The subproblems as well as the master problem we obtain in above decomposition are much smaller than the original 2L-VRP-TL, and therefore there might be hope that they can be solved to proven optimality in practice. However, all these are still NP-hard problems, and we pay the price of using a decomposition by usually having to solve many instances of the master and subproblems.

For generally improving scalability to larger instances, we can turn the exact BD approach into a faster approximate one by solving the subproblems and/or the master problem only approximately. When we terminate the MIP-solver on each of these in-stances early after reaching a solution with costs that are guaranteed to not exceed a specified optimality gap of $p\%$ and we obtain a feasible final solution, we can be sure that this solution's cost also does not exceed an optimal value by more than $p\%$.

While this might be a practical approach in some cases, the MIP-solver will often still require too much time to obtain approximate solutions with reasonable quality guaran-tees. In fact, experiments indicated in our scenario that only very moderate speedups could be achieved when allowing a gap of 5%. Suitable metaheuristics appear to be a more promising alternative.

6.1 Heuristic BD

We might consider virtually any well-working metaheuristic for the VRP with time-windows which is not too slow to approach our master and subproblem. For our proof-of-concept experiments here, we decided to apply the following previous work for the periodic VRP with time-windows [21]:

- One initial solution is created by Clarke and Wright's savings algorithm [22], which is adapted in a straight-forward way to only merge feasible routes w.r.t. the time limits.
- A set of n_{init} further, diverse initial solutions is derived by applying a randomized variant of the savings algorithm. The savings of combining two tours is accepted as the currently best savings if its value multiplied by a uniformly distributed random value within $[0.7, 1.3]$ is greater than the previously best known savings.
- The best initial solution undergoes *variable neighborhood descent* [23] using the following neighborhood structures in this order: *intra-route 2-opt*, *intra or-opt* (sequences of one, two, or three stations are moved to another position), and *inter-route 2-opt** (exchange of all feasible end-segments among two routes); for details see [21]. A first-improvement strategy is applied and the procedure only stops after reaching a locally optimal solution w.r.t. all these neighborhoods. Each candidate solution is checked for feasibility concerning the time limits and only feasible solutions are accepted.

In the BD, this metaheuristic can directly replace the exact resolution of the master and subproblems by the MIP-solver. However, we must take care in the bookkeeping of already known solutions $(t_s^k, x^k, c_s^k, \hat{t}_s^k)$ as they are not necessarily optimal anymore. On the one hand, a later identified solution for a time t_s^l may dominate earlier solutions $t_s^k < t_s^l$ even with lower cost, i.e., $c_s^k < c_s^l$. Thus, existing entries need to be verified and must possibly be removed together with the corresponding cuts. On the other hand, it may also happen that a newly found solution $(t_s^l, x^l, c_s^l, \hat{t}_s^l)$ has higher cost than an already known solution $(t_s^k, x^k, c_s^k, \hat{t}_s^k)$ with $\hat{t}_s^k \geq t_s^l$. In this case, no new violated cut can be derived, we may just store $(t_s^k + \varepsilon, x^l, c_s^l, \hat{t}_s^l)$ and the corresponding cut if $\hat{t}_s^l \geq \hat{t}_s^k$ for possible future use.

7 Computational Experiments

We compare the performance of directly solving the MIP model (7)–(11) for 2L-VRP-TL, the MIP-based exact BD approach, and two variants of the heuristic BD on a set of synthetic Euclidean instances and instances based on the TSPlib[1].

All algorithms have been implemented with GCC 4.6. Each test run was performed on a single core of an Intel Xeon E5540 machine with 2.53 GHz. CPLEX version 12.1 was used for solving the MIPs.

---

[1] https://www.iwr.uni-heidelberg.de/groups/comopt/software/ TSPLIB95

7.1 Instances

The synthetic Euclidean instances can be divided into four subtypes:

- instances where the first-level VRP and the second-level VRPs are equally large, i.e. $|V_0| = |V_s|$, $\forall s \in V_0'$,
- instances with a larger first-level VRP, i.e. $|V_0| > |V_s|$, $\forall s \in V_0'$,
- instances with larger second-level VRPs, i.e. $|V_0| < |V_s|$, $\forall s \in V_0'$, and
- instances where each of the second-level VRPs has different size, i.e. $|V_s| = \lceil 0.5|V_0| \rceil + 1, \ldots, \lfloor 1.5|V_0| \rfloor$.

For the first-level VRP, satellites are randomly placed on a 201×201 grid with the depot node being located at the center. Each second-level VRP is constructed essentially in the same way considering a separate grid of the same size: The satellite is assumed to be at the center, and all customers are placed randomly at the grid. Traveling times are rounded Euclidean distances, and traveling costs are derived from these times by adding uniform random perturbations of 20%. Demands are chosen randomly from $\{1, \ldots, 100\}$. The vehicle capacity and the global time limit were selected manually in a way that the instances are non-trivial.

For the TSPlib instances we applied a clustering that roughly simulates the process of opening satellite depots in real-world. Given the basic nodes which represent customers, we added satellites manually at plausible locations and then assigned each customer node to the closest satellite. As a result, the sizes of the second-level VRPs differ to a certain degree. Customer demands are chosen randomly from $\{1, \ldots, 10\}$ while traveling times and costs, vehicle capacity and the global time limit are determined in the same way as above. All instances are available online[2].

7.2 Results

Tables 1 and 2 compare the following algorithm variants: directly solving the MIP model (7)–(11) ("pure MIP"), the MIP-based exact BD, the BD variant where the subproblems are solved heuristically, and the fully heuristic BD variant where the master problem as well as the subproblems are solved heuristically. Synthetic instances are specified by the size of the master problem ($|V_0|$) and the size of the subproblems ($|V_s|$). For the TSPlib problems we only list $|V_0|$ since the subproblems have different sizes.

Table 1 shows the objective values of final solutions and the required CPU times. Best values are printed bold for each instance. For the exact BD variants, we list the gaps between lower and upper bounds after reaching the time limit of one hour. For the variants where we use heuristics, 30 independent runs were performed in order to obtain average objective values of final solutions and standard deviations. The time limit was set to 10min. Table 2 displays for the BD variants further information on the number of added Benders' cuts and the number of times the master problem is (re-)solved.

First of all, we observe that the pure MIP approach is only viable for small instances where the size of the sub-VRPs is at most 15. For larger instances the gaps are soon too large for the solutions to be meaningful. The exact MIP-based BD performs better on

---

[2] https://www.ads.tuwien.ac.at/w/Research/Problem_Instances

Table 1. Solution qualities and CPU-times of different algorithm variants

Synthetic instances

$|V_0|$	$|V_s|$	pure MIP			BD using MIP			heur. BD for subp.			fully heuristic BD		
		obj	gap[%]	time[s]	obj	gap[%]	time[s]	$\overline{obj}$	sd	time[s]	$\overline{obj}$	sd	time[s]
5	5	**2844.0**	0.0	0.0	**2844.0**	0.0	0.1	**2844.0**	0.0	0.0	**2844.0**	0.0	0.0
5	3	**1450.0**	0.0	0.0	**1450.0**	0.0	0.0	**1450.0**	0.0	0.0	**1450.0**	0.0	0.0
3	5	**1113.0**	0.0	0.0	**1113.0**	0.0	0.1	**1113.0**	0.0	0.0	**1113.0**	0.0	0.0
5	4...7	**2514.0**	0.0	0.0	**2514.0**	0.0	0.1	**2514.0**	0.0	0.0	**2514.0**	0.0	0.0
9	9	8250.0	0.8	0.2	**8235.0**	0.0	0.3	**8235.0**	0.0	0.1	**8235.0**	0.0	0.1
9	5	**5304.0**	0.0	0.0	**5304.0**	0.0	0.1	**5304.0**	0.0	0.0	**5304.0**	0.0	0.0
9	9	5376.0	0.2	0.1	**5376.0**	0.0	0.1	5400.0	0.0	0.0	5400.0	0.0	0.0
9	6...13	9337.0	0.2	1.9	**9337.0**	0.0	0.5	**9337.0**	0.0	0.2	9359.0	0.0	0.2
15	15	18367.0	4.7	600.0	**17518.0**	0.0	49.0	17704.8	34.4	7.7	17709.0	24.3	1.7
15	9	11270.0	0.5	5.2	**11268.0**	0.0	4.3	11271.2	12.0	5.4	11265.4	10.8	0.5
8	15	9307.0	0.6	600.0	**9093.0**	0.0	41.8	9096.7	6.5	0.6	9094.9	4.3	0.6
15	9...22	21506.0	2.5	900.0	22218.0	15.0	24.7	21148.5	49.4	2.3	**21147.6**	54.3	1.6
25	25	40708.0	22.1	3600.0	42299.0	15.0	3590.0	25203.2	24.9	577.0	40000.4	63.2	26.0
25	14	23932.0	14.9	3600.0	25817.0	15.0	3600.0	**18788.4**	23.1	76.4	**23384.0**	34.8	11.3
13	25	18880.0	21.6	3600.0	-	-	3600.0	39138.2	332.4	600.0	18793.9	29.6	10.5
25	14...37	38881.0	23.5	3600.0	-	-	3600.0	-	-	600.0	**38173.0**	63.3	38.9
35	35	70467.0	36.0	3600.0	42343.0	15.0	236.4	30474.6	133.9	600.0	**60630.8**	84.8	136.3
35	19	39536.0	19.2	3600.0	-	-	3600.0	-	-	590.0	38517.2	57.2	40.9
18	35	31275.0	28.5	3600.0	-	-	3600.0	-	-	600.0	**30061.0**	63.0	56.7
35	19...52	69238.0	34.7	3600.0	-	-	3600.0	-	-	600.0	**61041.1**	87.9	236.8
50	50	234161.0	69.2	3600.0	-	-	3600.0	-	-	600.0	**100823.7**	356.3	600.0
50	27	67746.0	26.9	3600.0	-	-	3600.0	-	-	600.0	62687.1	81.0	316.3
26	50	76159.0	51.4	3600.0	-	-	3600.0	-	-	600.0	**52019.9**	60.8	331.1
50	27...75	226549.0	67.3	3600.0	-	-	3600.0	-	-	600.0	**105743.4**	702.4	600.0

TSPlib instances

Name	$|V_0|$	pure MIP			BD using MIP			heur. BD for subp.			fully heuristic BD		
		obj	gap[%]	time[s]	obj	gap[%]	time[s]	$\overline{obj}$	sd	time[s]	$\overline{obj}$	sd	time[s]
a280	20	6004.0	23.3	3600.0	**5762.0**	0.0	1176.1	5798.2	9.6	33.6	5798.5	11.5	6.5
berlin52	6	**13784.0**	3.1	3600.0	**13784.0**	0.0	12.8	13784.8	2.0		13786.1	4.8	0.2
bier127	13	204856.0	7.1	3600.0	**204361.0**	0.0	180.5	204464.3	94.1	8.5	204407.8	163.9	1.0
ch130	11	13727.0	7.8	3600.0	**13704.0**	0.0	558.6	13747.8	19.1	1.1	13751.2	24.8	1.1
pr1002	16	2347399.0	80.3	3600.0	-	-	3600.0	18158.8	39.0	389.1	**18153.7**	41.1	389.9
rat783	13	64684.0	81.2	3600.0	-	-	3600.0	**715598.6**	1261.9	265.6	716024.0	1242.4	249.3

Table 2. Numbers of generated cuts and master problem resolves of the BD variants

Synthetic instances							
		BD using MIP		heur. BD for subp.		fully heuristic BD	
$|V_0|$	$|V_s|$	#cuts	#resolves	#cuts	#resolves	#cuts	#resolves
5	5	4.0	3.0	4.0	3.0	4.0	3.0
5	3	1.0	2.0	1.0	2.0	1.0	2.0
3	5	3.0	3.0	3.0	3.0	3.0	3.0
5	4...7	3.0	2.0	3.0	2.0	3.0	2.0
9	9	10.0	3.0	3.0	2.0	7.0	3.0
9	5	5.0	3.0	5.0	3.0	4.0	3.0
5	9	3.0	2.0	3.0	2.0	3.0	2.0
9	6...13	10.0	4.0	9.0	3.0	9.4	3.3
15	15	32.0	7.0	30.3	7.1	27.6	6.6
15	9	34.0	6.0	34.9	7.0	22.0	4.1
8	15	12.0	5.0	10.2	5.0	10.2	5.0
15	9...22	13.0	2.0	28.0	6.8	25.8	6.3
25	25	16.0	2.0	-	-	87.9	11.5
25	14	12.0	2.0	31.7	1.0	104.9	14.5
13	25	-	-	41.9	8.5	39.6	7.9
25	14...37	-	-	47.4	2.1	105.4	12.9
35	35	-	-	-	-	125.9	12.3
35	19	17.0	2.0	29.0	3.2	165.4	13.2
18	35	-	-	-	-	56.2	11.1
35	19...52	-	-	-	-	182.8	16.9
50	50	-	-	-	-	138.4	8.4
50	27	-	-	-	-	251.4	22.3
26	50	-	-	-	-	76.9	12.3
50	27...75	-	-	-	-	113.4	4.1

instances with sub-VRPs with up to 15 nodes. However, on larger instances it is often not able to solve all subproblems of the first major iteration in time, and we therefore do not get any feasible solution for the master problem. For instances with sub-VRPs of size 25 or more, this even holds when terminating CPLEX early with an optimality gap limit of 15%. We remark that this rather bad behavior is particularly due to the relatively weak Miller-Tucker-Zemlin formulation, and one can expect to improve the situation by using a more state-of-the-art exact VRP solver. When using the heuristic approach for the BD subproblems, they are solved in a relatively short time to very reasonable quality so that a feasible solution to the master problem can be obtained most of the time. However, on the larger instances it proves to be difficult nonetheless to solve the master problem via MIP and in some cases it was not possible to solve it even once within the time limit. The fully heuristic BD works well on small instances where optimal solutions are reliably reached in short times. On larger instances it has excellent scalability and produces by far the best results.

Comparing the different subtypes of synthetic instances, we clearly see that the most challenging ones are those where the master problem and the subproblems are equally large. The pure MIP approach and the exact BD approach are able to solve instances with small subproblems in comparison faster since they can concentrate on the master problem. This is expressed by the low number of Benders' cuts that are added and thus the low number of times that the master problem has to be (re-)solved, see Table 2. Instances where subproblems have different sizes are usually also easier to solve. The reason here is that the master problem often is easier: In the first-level VRP, the satellites of larger subproblems typically need to be visited earlier than satellites of smaller subproblems. In Table 2 we also observe that for larger instances, the fully heuristic BD variant is able to perform much more iterations, i.e., more Bender's cuts are added and the master problem is more often resolved within the same time limit.

8 Conclusions

Logic-based BD is a promising extension of classical BD to tackle far more problems from practice because subproblems are not restricted to LPs anymore. Its application to our 2L-VRP-TL is relatively intuitive and Benders' cuts can be derived from primal subproblem solutions and its inference dual in a rather straight-forward way.

Applying the logic-based BD with CPLEX for exactly solving all master and subproblem instances turned out to be beneficial for some mid-size instances in comparison to solving the original MIP formulation directly. Problematic, however, are the long running times for solving the Benders' subproblems, preventing the approach from finding feasible solutions to larger 2L-VRP-TL instances at all. By just aiming for good approximate solutions and using a variant of variable neighborhood search to heuristically solve the master and subproblem instances, we could dramatically improve the scalability and obtain by far the best results on the considered instances.

More generally, it was shown that logic-based BD may be a fruitful framework also for metaheuristics. While previous work already documented the usefulness of metaheuristics for approximately solving the master problem in classic BD approaches, our work goes beyond and applies a metaheuristic especially to more complex subproblems (and their inference duals) as they appear in logic-based BD. The general technique seems to be promising also for other classes of problems and deserves further research.

Our implementation for 2L-VRP-TL only is a first proof-of-concept. It is obvious that it can be improved on the one hand by utilizing a tighter MIP-formulation for $VRP(G_s)$, e.g., based on multi-commodity flows, or even a more sophisticated branch-and-cut. On the other hand a more advanced metaheuristic may also be chosen for $VRP(G_s)$. The principles of the logic-based BD and the combination with the metaheuristic, however, stay the same.

Future work should in particular investigate a combined application of heuristic and exact methods for solving the Benders' subproblems. For example, one can first solve the subproblems heuristically yielding approximate Benders' cuts and a heuristic solution quickly. In a second phase, the existing Benders' cuts are iteratively validated by solving corresponding subproblems exactly, exploiting the already known heuristic solutions. Possibly found improved subproblem solutions yield new exact Benders' cuts that replace the dominated heuristic cuts. When resolving the master problem and validating all Benders' cuts in this way, an exact solution is obtained in the end.

References

1. Benders, J.F.: Partitioning procedures for solving mixed-variables programming problems. Numerische Mathematik 4, 238–252 (1962)
2. Geoffrion, A.M.: Generalized Benders decomposition. Journal of Optimization Theory and Applications 10(4), 237–260 (1972)

3. Hooker, J.N., Ottosson, G.: Logic-based benders decomposition. Mathematical Programming 96, 33–60 (2003)
4. Hooker, J.N.: Planning and scheduling by logic-based Benders decomposition. Operations Research 55(3), 588–602 (2007)
5. Poojari, C.A., Beasley, J.E.: Improving Benders decomposition using a genetic algorithm. European Journal of Operational Research 199(1), 89–97 (2009)
6. Lai, M.C., Sohn, H.S., Tseng, T.L., Chiang, C.: A hybrid algorithm for capacitated plant location problem. Expert Systems with Applications 37(12), 8599–8605 (2010)
7. Lai, M.C., Sohn, H.S.: Using a genetic algorithm to solve the Benders master problem for capacitated plant location. In: Gao, S. (ed.) Bio-Inspired Computational Algorithms and Their Applications, pp. 405–420. InTech (2012)
8. Lai, M.C., Sohn, H.S., Tseng, T.L., Bricker, D.L.: A hybrid Benders/genetic algorithm for vehicle routing and scheduling problem. International Journal of Industrial Engineering 19(1), 33–46 (2012)
9. Rei, W., Cordeau, J.F., Gendreau, M., Soriano, P.: Accelerating Benders decomposition by local branching. INFORMS Journal on Computing 21(2), 333–345 (2008)
10. Zakeri, G., Philpott, A.B., Ryan, D.M.: Inexact cuts in Benders decomposition. SIAM Journal on Optimization 10(3), 643–657 (1999)
11. Perboli, G., Tadei, R., Vigo, D.: The two-echelon capacitated vehicle routing problem: Models and math-based heuristics. Transportation Science 45(3), 364–380 (2011)
12. Hemmelmayr, V.C., Cordeau, J.F., Crainic, T.G.: An adaptive large neighborhood search heuristic for two-echelon vehicle routing problems arising in city logistics. Computers and Operations Research 39(12), 3215–3228 (2012)
13. Schwengerer, M., Pirkwieser, S., Raidl, G.R.: A variable neighborhood search approach for the two-echelon location-routing problem. In: Hao, J.-K., Middendorf, M. (eds.) EvoCOP 2012. LNCS, vol. 7245, pp. 13–24. Springer, Heidelberg (2012)
14. Crainic, T.G., Mancini, S., Perboli, G., Tadei, R.: Multi-start heuristics for the two-echelon vehicle routing problem. In: Hao, J.-K. (ed.) EvoCOP 2011. LNCS, vol. 6622, pp. 179–190. Springer, Heidelberg (2011)
15. Jacobsen, S.K., Madsen, O.B.G.: A comparative study of heuristics for a two-level routing-location problem. European Journal of Operational Research 5(6), 378–387 (1980)
16. Boccia, M., Crainic, T., Sforza, A., Sterle, C.: A metaheuristic for a two echelon location-routing problem. In: Festa, P. (ed.) SEA 2010. LNCS, vol. 6049, pp. 288–301. Springer, Heidelberg (2010)
17. Boccia, M., Crainic, T.G., Sforza, A., Sterle, C.: Location-routing models for designing a two-echelon freight distribution system. Technical Report CIRRELT-2011-06, University of Montreal (2011)
18. Fisher, M.L., Jaikumar, R.: A generalized assignment heuristic for vehicle routing. Networks 11(2), 109–124 (1981)
19. Boschetti, M., Maniezzo, V.: Benders decomposition, Lagrangian relaxation and metaheuristic design. Journal of Heuristics 15, 283–312 (2009)
20. Yaman, H.: Formulations and valid inequalities for the heterogeneous vehicle routing problem. Mathematical Programming 106(2), 365–390 (2006)
21. Pirkwieser, S., Raidl, G.R.: A variable neighborhood search for the periodic vehicle routing problem with time windows. In: Prodhon, C., et al. (eds.) Proceedings of the 9th EU/MEeting on Metaheuristics for Logistics and Vehicle Routing, Troyes, France (2008)
22. Clarke, G., Wright, J.W.: Scheduling of vehicles from a central depot to a number of delivery points. Operations Research 12(4), 568–581 (1964)
23. Mladenović, N., Hansen, P.: Variable neighborhood search. Computers and Operations Research 24(11), 1097–1100 (1997)

Author Index